Down from the Mountaintop

MELISSA WALKER

Down from the Mountaintop

BLACK WOMEN'S NOVELS IN THE WAKE OF THE CIVIL RIGHTS MOVEMENT, 1966–1989

YALE UNIVERSITY PRESS

NEW HAVEN AND LONDON

Designed by Nancy Ovedovitz and set in Janson type by G&S Typesetters. Printed in the United States of America by BookCrafters, Inc., Chelsea, Michigan.

Library of Congress Cataloging-in-Publication Data
Walker, Melissa.
Down from the mountaintop: black women's novels in the wake of the civil rights movement, 1966–1989 / Melissa Walker.
p. cm.
Includes bibliographical references and index.
ISBN 0-300-04855-6 (alk. paper)
1. American fiction—Afro-American authors—History and criticism. 2. Women and literature—United States—History—20th century. 3. American fiction—Women authors—History and criticism. 4. American fiction—20th century—History and criticism. 5. Civil rights movements in literature. 6. Afro-American women in literature. 7. Afro-Americans in literature. I. Title.
PS374.N4W35 1991
813'.54099287—dc20 90-40653
CIP

The paper in this book meets the guidelines for permanence and durability of the Committee on Production Guidelines for Book Longevity of the Council on Library Resources.

10 9 8 7 6 5 4 3 2

CONTENTS

ACKNOWLEDGMENTS

This book owes much to many: novelists, civil rights activists, librarians, scholars, students, editors, friends, and family. Two writers whose generosity to me influenced the way this book developed are Maya Angelou, who spent a day responding to my questions, and Rosellen Brown, whose novel *Civil Wars* and enthusiastic and informed conversations made a difference in my thinking about the interrelation of private and public lives.

Librarians who graciously assisted me are Janet Sims-Wood at the Moreland-Spingarn Research Center and Louise Cook, then at the Martin Luther King, Jr., Center for Nonviolent Social Change. Sue Thrasher gave me access to the files of the Highlander Folk School and provided me with information that helped me formulate the early steps of the project.

Novelists who answered my questions about their work in public conferences or private conversations include Toni Cade Bambara, Alice Childress, Rosa Guy, Kristin Hunter, Louise Meriwether, Toni Morrison, Alice Walker, and Margaret Walker.

I have been fortunate in friends who have had faith that I would do this work, to use Toni Morrison's words, "appropriately and well": Roberta Bondi, Sandra Deer, Jingle Davis, Sue Dennard, Alice Goldsmith, Carol Light, and Marjorie Shostak. Marilyn Montgomery, Virginia Ross, and Eliza Davis proofread large portions of the manuscript on the spur of the moment. I also wish to thank Bill Andrews, Tina McElroy Ansa, Kimberly Bentson, Blanche Gelfant, Barbara Hardy, Paul Hunter, Melvin Konner,

George Levine, Barry Wade, Cornel West, and Jim Yerkes for facilitating the progress of this work.

When Marymal Dryden invited me to serve on the steering committee of a national conference on "Women in the Civil Rights Movement," she provided me with an opportunity to meet many scholars of the civil rights movement, as well as activists who substantially affected the progress of the movement. Especially helpful were Dr. Broadus Butler, Clay Carson, Vicki Crawford, Jacqueline Anne Rouse, and Barbara A. Woods. Particularly inspiring were activists Anne Braden, Virginia Durr, Rosa Parks, Mamie Till Mobley, Bernice Johnson Reagon, and Miles Horton. The work of critics and scholars of African-American literature is acknowledged in the introduction and notes of the text.

Students in my African-American literature classes whose challenging questions helped refine my thinking include Hope Kornegay, Karla Small, and Suzanne Yaeger.

Jerome Beaty's contribution to this book has been extensive and substantive. His careful reading of early and late drafts influenced significantly the character and shape of the book. I am also deeply indebted to Elizabeth Fox-Genovese for the invaluable suggestions that followed her unflinching scrutiny of the manuscript.

I am profoundly appreciative of my editor Ellen Graham at the Press for the care and attention she gave this work and for her timely guidance and encouragement in seeing it through to completion. For her splendid work as manuscript editor, I thank Stacey Mandelbaum.

For meticulous reading of draft after draft, I thank my husband, Jerome Walker; for diligent editing of crucial portions of the final manuscript, I thank my son, Richard Walker. I am especially grateful to my daughter, Laura Walker, for her ongoing affirmation of this project. For their enduring confidence in the value of my work, I thank my parents, Richard and Pearl Graves, and my sister, Lucy Hughes.

INTRODUCTION

The civil rights movement is remembered by many as a series of supreme moments followed by chastening downfalls. The mountaintop experiences of *Brown v. Board of Education,* the Montgomery bus boycott, the victories of the sit-ins and freedom rides, the 1963 March on Washington, and the final success of the Selma protests were regularly punctuated by bombings, lynchings, assassinations, and riots. The brief euphoria following the *Brown* decision, when many believed that integration and full equality would be achieved by 1963, soon gave way to the chilling apprehension that the road to the promised land would be strewn with the bodies of martyrs.

Today the names of only a few of the victims are remembered—Emmett Till, Medgar Evers, James Chaney, Andrew Goodman, Michael Schwerner, and, of course, Martin Luther King, Jr. We will never know exactly how many others died as a consequence of the backlash that succeeded the first effective challenges to Jim Crow. Even more difficult to assess are the psychological wounds suffered by both activists and innocent victims of racist reprisals. What became apparent to the immediate survivors of the movement was that the view of the promised land would be increasingly obscured by the fallout from repeated acts of violence. In early 1990, as I write these words, the smoke has barely cleared from the latest explosions of letter bombs seemingly intended to terrorize those who fight for civil rights.

Since the early 1960s there have been growing antagonisms between various factions of the activist community: discord among blacks over issues of militancy and nonviolence, the role of whites in the movement, and the relative merits of economic and political goals; conflicts between black women crying sexism and black men claiming slander; and the estrangement of African Americans from some of their strongest allies in the Jewish community. Since the mid-sixties, the concentrated vision of racial harmony has been diluted by the proliferation of other causes—including the movement to end the Vietnam War, and the women's, the antinuclear, and the environmental movements—as well as efforts to expand the call for civil rights to one for human rights for all groups and individuals who have been victims of discrimination or exploitation.

During the years between *Brown v. Board of Education* in 1954 and the passage of the Voting Rights Act in 1965, many African Americans who favored integration and assimilation believed that their prolonged struggle for social justice would soon be won and that the "beloved community" envisioned by King and others would presently be a reality. During the last years of the sixties that belief was gradually eroded, and by the time King was assassinated in April 1968 most activists and many ordinary folk who had been inspired by the movement were floundering, wondering what had happened to the dream. Others who had staked their futures on the separatist movement and a commitment to black political power watched their ambitions go up in the flames of urban riots. Both groups were soon asking the question—Where Do We Go from Here?—that King had used as the title for his last book. That question and its corollaries—Where Are We? and Where Have We Been?—provided the impetus for a larger cultural inquiry.

A significant number of African-American women writers took part in that inquiry in the immediate aftermath of the civil rights movement. Producing a steady stream of notable fictional narratives ranging from Margaret Walker's *Jubilee* to Alice Walker's *The Temple of My Familiar*, they have addressed these and other relevant questions about the past, present, and future of the African-American community. The way they have focused and composed their narratives relates directly to the civil rights movement—its issues, events, and consequences.

The subject of the eighteen novels that I consider in this book is "where we are"; the issue they explore is "where we are going"; and both are manifested in the authors' particular visions of "where we have been," as revealed in the events they focus on in the journey upward from slavery: rebellions, flights to freedom, preparation for legal action, outbursts of cultural nationalism, nonviolent protests, and development of a grass roots base for political power. Taken as a group, these novels contain details from many chapters of black history: they evoke for readers the holds of slave ships, the speaker's corner in Harlem, and sharecroppers' shacks, but also the polished tables in the carefully decorated homes of the black bourgeoisie.

From *Jubilee* to *The Temple of My Familiar*, events in the private lives of fictional characters are narratively linked to particular episodes in the struggle for racial justice. The long list of such historical allusions includes operations of the Underground Railway in the antebellum period; the Fugitive Slave Law of 1850; the Emancipation Proclamation of 1863; the constitutional and political achievements in the 1865–77 Reconstruction and the terrorist reprisals of the Ku Klux Klan (KKK) during the same period; the withdrawal of federal troops and redemption of power by southern conservatives after the Hayes Compromise in 1877; the legalization of segregation with the separate-but-equal doctrine in *Plessy v. Ferguson* in 1896; the development of factions within the black community roughly divided into followers of Booker T. Washington advocating accommodation to white supremacy and those influenced by W. E. B. Du Bois favoring legal action to achieve civil rights; Marcus Garvey's launching in 1916 of the Universal Negro Improvement Association (the first mass movement of African Americans); the public outcry following the arrest of nine young black men in Scottsboro, Alabama, in 1931; Roosevelt's executive order banning discrimination in defense industries and government employment in 1941 and Truman's desegregation of the armed forces in 1948; the founding of the Congress of Racial Equality (CORE) in 1942; the desegregation of public schools in *Brown v. Board of Education* in 1954; the public protest following the murder of Emmett Till in 1955; the Montgomery Boycott of 1955–56; the founding of the Southern Christian Leadership Conference (SCLC) in 1957 under the leadership of Martin Luther King, Jr., and of the Student Nonviolent Coordinating Committee (SNCC) in 1960 under the auspices of Ella Baker; and finally the day-to-day activities of those engaged in the movement in its heyday. While these and other specific events of the more than a century of black protests and rebellions are an overt and integral part of the novels considered here, other elements of that long process are embedded in the narratives in ways that invite readers to seek out the connections between private lives and their public context.[1]

Just as the civil rights movement facilitated the entrance of many African Americans into the middle class, a journey that a character in Toni

Morrison's *The Bluest Eye* refers to as moving from the hem "into the major fold of the garment of life" (18), so has it accelerated the current outpouring of remarkable novels by African-American women. The degree to which a novelist's characters are comfortable with their place in the middle class says something about that novelist's own attitude toward the value of the journey that brought them there. From Claudia in *The Bluest Eye* to Jadine in *Tar Baby*, Morrison's middle-class black characters seem tortured by the price they have paid for their social status (or assimilation into mainstream integrated society). On the other hand, Ntozake Shange's Betsey Brown looks out at the suffering of others with the comfortable confidence of noblesse oblige, while Alice Walker's Shug Avery does not seem remotely aware of the considerable poverty that plagued the rural South during the years when she was making it as a famous blues singer.

All the novels treated here are rooted in history and in the culture and community of African-American life that can be traced to the experience of slavery. Each is set in a specific geographic and temporal—and, in that sense, historical—context. Not all of the novels written by black women since the mid-sixties seem relevant to such a study. Excluded are science fiction novels, those with major characters who have Caribbean rather than Southern roots, those with little or no reference to African-American culture and community, and narratives focusing primarily on personal and sexual themes divorced from larger social issues. Novels by Octavia Butler, Pauli Marshall, Gayle Jones, and Charlene Hatcher Polite are examples. Autobiographies such as those by Pauli Murray, Anne Moody, Maya Angelou, Mary E. Mebane, and Marita Golden are excluded on generic grounds. This book does not, however, treat all recent novels by black women that relate to the movement: Gloria Naylor's novels contain characters who were movement activists (Kiswana Browne in *Brewster Place*), admirers of Malcolm X (Lester Tilson in *Linden Hills*), or victims of racist employment practices (Ophelia in *Mama Day*), but their emphasis is elsewhere—on the internal conflicts of black communities. Very recent novels published since this study was completed are Barbara Chase-Riboud's *Echo of Lions*, Tina McElroy Ansa's *Baby in the Family*, and Marita Golden's *Long Distance Life*.

The eighteen novels considered here are grouped in chapters according to their historical setting—where we have been. Within the chapters they are arranged according to date of publication—where we are (or were at the time of publication), which is often as significant in terms of the movement and its aftermath as the historical setting itself.

Chapter 1 treats three novels set during the period of slavery and reconstruction. Margaret Walker's *Jubilee* (1966), with its outright celebration of progress in the struggle for racial justice and harmony within the black community, set in the years immediately preceding, during, and following the Civil War, is clearly a product of the mountaintop visions of the mid-sixties; Sherley Anne Williams's *Dessa Rose*, published twenty years later in 1986, extols small private victories in the 1847 world that placed little value on relieving the suffering of slaves. Williams's novel is quite at home in the political climate of the 1980s, when victories for African Americans were private ones for those within or entering the middle class rather than public ones affecting the growing underclass. Published a year later in 1987, Toni Morrison's *Beloved* turns attention to former slaves and to the most devastating aspects of their lives in bondage, suggesting the process that would lead some up from slavery and eventually into the middle class and others to permanent social marginality and to early death. As in her previous work, Toni Morrison is inclusive, relevant—and ambivalent.

In chapter 2, I examine Toni Morrison's *The Bluest Eye* (1970) and Alice Walker's *The Color Purple* (1982), both set chiefly in the period between the world wars. Written in the last years of the 1960s, as young activists advocating black power challenged the old guard still urging accommodation to the white establishment, *The Bluest Eye* explores the traumatic consequences of the cultural glorification of the middle-class white aesthetic on one child whose poverty and infatuation with whiteness lead to madness and on another who dismisses her less fortunate friend in her scramble to join the middle class. Published in 1970 at the height of the "black is beautiful" movement, Morrison's novel reinforces a major concern of the black community at that time and examines the dangers of the more subtle manifestations of white supremacy. *The Bluest Eye* is narratively linked to the civil rights period through the voice of Claudia MacTeer, one of its narrators, who looks back and regrets that she has achieved her own place in society by buying into the white value system and stepping over those who were destroyed by it, thus inviting readers to consider their role in the very recently altered world. *The Color Purple* was published twelve years later, when American culture was dominated by a complacent acceptance of free market rhetoric and a concern with feminist issues. Lacking a narrative link to the present, this novel allows readers to dwell in its once-upon-a-time world and to avoid considering the direction that racism and its consequences were taking in the early 1980s. While this best-seller includes scenes that dramatize the endemic racism of the first four decades of the century, Celie's success in business reinforces America's love affair with entrepreneurial capitalism. In the final scene of the novel, the once estranged or geographically separated characters are united in an all black community that would hardly threaten even middle-class readers comfortably settled in their segregated neighborhoods.

In chapter 3, I discuss three novels set in Harlem: Louise Meriwether's *Daddy Was a Number Runner* (1970), Alice Childress's *A Short Walk* (1979), and Rosa Guy's *A Measure of Time* (1983). Composed in the last years of the sixties, published the same year as *The Bluest Eye*, and set in the mid-1930s, *Daddy* relates the private life of Francie Coffin to the economic realities of the Depression, as well as to the larger public arena. Meriwether connects her story to the time of its publication by including scenes in which characters imagine and prepare for the day when changes will come and by suggesting how the past is the precondition for the present. The action of Alice Childress's *A Short Walk*, spanning the first half of the twentieth century, also commemorates the lives of those who paved the way for the civil rights movement, concluding with the first significant victory for those seeking racial equality—the desegregation of the armed forces. Celebrating the same period is Rosa Guy's *A Measure of Time*, which follows the characters' struggles with Jim Crow all the way to the Montgomery bus boycott. By the end of the seventies, when Childress's novel was published, the failure of the civil rights movement to achieve its goals of educational, economic, and social equality was generally apparent. Four years later, when Guy's novel appeared, the dream of a fully integrated society seemed to belong to the realm of fiction rather than social reality. By going back to that earlier period when the foundations were being laid for the movement and its successes, these two novels celebrate the achievements of those preparatory years while inviting readers to consider what the future for African Americans might be in a society increasingly bent on pursuing private concerns.

Chapter 4 considers two novels that treat the private struggles of characters living in isolated, segregated communities from the end of World War I through the beginning of the civil rights movement. Alice Walker's first novel, *The Third Life of Grange Copeland*, like *The Bluest Eye* and *Daddy Was a Number Runner*, was composed during the height of the movement and published in 1970. Toni Morrison's *Sula*, also an outgrowth of the final years of the movement, was published in 1973. Though *Grange* and *Sula* are set in the same years and are products of approximately the same period, *Grange* looks forward to possibility and *Sula* looks backward to loss. Both, however, conclude with a character's private confrontation with the promises of the civil rights movement.

The four novels discussed in chapter 5 are set primarily in the peak years of the movement. Key scenes of Toni Morrison's *Song of Solomon* converge with crucial episodes in the movement. Published in 1977, some fourteen years after the mountaintop high of the 1963 March on Washington, the climactic scenes of *Solomon* take place during the same weeks as the march and its immediate aftermath. Milkman's ambiguous perspective from the top of a precipice at the end of the novel invites consideration of the equally precarious postures of the black community and its visionaries, many of them either dead, retired, or living in exile by the end of 1977. Milkman's quest is less ambiguous than his fate, suggesting an alternative to a self-indulgent existence isolated from the life of the community. Kristin Hunter's *The Lakestown Rebellion* (1978) is set in the summer of 1965, following the Selma marches and during the final debates leading to the passage of the Voting Rights Act early in August. A story of local activists who successfully use the methods of the nonviolent movement to defeat the forces threatening their community, *Lakestown* celebrates the movement during the last days when an unqualified spirit of jubilation could prevail. Soon the fragile unity of the black community, nurtured for so long by the spirit of nonviolence and apparent victories, would be fractured by the demands of militants like Stokely Carmichael advocating "black power." Soon the heady spirit of mass protests would be replaced for some by the tedious work of registering voters to build a political base for the black community. Others would abandon activism, purportedly seeking to break legal, economic, and social barriers through their own personal successes. Ntozake Shange's *Sassafrass, Cypress, and Indigo* (1982), set during the sixties, and *Betsey Brown* (1985), which takes place in St. Louis in 1959, both gloss over enduring conditions that have plagued the black community long after the praises of movement leaders had begun to ring hollow. And both affirm the quest for private fulfillment in ways consistent with the 1980s pursuit of personal satisfaction to the exclusion of public commitments.

Chapter 6 focuses on three novels set in the post-movement period following the brief Second Reconstruction, when all three branches of the government cooperated to assure full rights for African Americans. The primary action of Alice Walker's *Meridian* (1976), Toni Cade Bambara's *The Salt Eaters* (1980), and Toni Morrison's *Tar Baby* (1981) takes place in the mid or late 1970s when the sweeping reforms of the 1960s reconstruction were being undermined by the "benign neglect" and the calculated retreat of the Nixon years, the ineffectiveness of Carter's civil rights policies, and the dilution of civil rights for blacks by his well-intentioned call for human rights for everyone, everywhere. The three main characters of *Meridian*, all activists during the movement's heyday, are plagued by the personal repercussions of their public commitments during the early sixties. Rather than a series of mountaintop visions leading to a clear mandate for the future, the movement toward freedom was chaotic, conflict ridden, and personally devastating. The characters in *The Salt Eaters* contend with the painful memories of their days in the movement and with the increasing demands of proliferating causes, which both expand and dilute the civil rights agenda—causes ranging from the fight for women's rights to the prevention of nuclear disaster. *Tar Baby* explores the unresolved tensions between assimilated blacks and separatists, as well the ongoing conflicts between blacks and whites.

The characters of Alice Walker's *The Temple of My Familiar* (1989), discussed in the Afterword, are still grappling in the late eighties with the question raised by Martin Luther King, Jr., in 1968. For activists like Fanny, answering "where do we go from here" requires confronting where she is and where she has been, questions that have informed the agenda of novels considered in this study.

While I will explore the ways a novel responds to and reflects prevailing values in the culture, I do not mean that at a particular time there is a monolithic zeitgeist determining the kinds of narratives writers produce and readers affirm. Rather, I suggest that novelists might write directly into the prevailing cultural values or into the counterforces challenging those values, but that in either case the spirits of the time are at work shaping their narratives. That significant elements of *The Color Purple* are compatible with the dominant ideology of the early eighties, for example, does not mean that all successful novels published at that time were necessarily speaking into or out of the identical cultural context. Toni Morrison's *Tar Baby*, discussed in chapter 6, calls into question the validity of the capitalist enterprises that *The Color Purple* seems implicitly to endorse. Walker's novel, on the other hand, reinforces the position of African Americans who maintain that economic independence is more liberating than privileges granted by a white power structure.

That both *Tar Baby* and *The Color Purple* were well-received and widely read novels at the beginning of the 1980s suggests that some readers may have responded to both Morrison's consideration of the damage inflicted by an exploitive system and Walker's evocation of the allure of the personal benefits of free enterprise. The same readers who thrilled to Shug Avery's appeal to self-indulgence in *The Color Purple* might well have been moved to applaud its opposite by Morrison's tough-minded integrity. Indeed this ambivalence itself was virtually a characteristic of the eighties. Nevertheless, it was *The Color Purple* that became a runaway best-seller, won a Pulitzer Prize, and reached an even larger audience in a much discussed movie. Similarly, though most Americans express concern about the destruction of the rain forest, an overt concern in *Tar Baby*, a substantial majority of voters in the eighties seemed to endorse the extravagant, short-sighted, and self-indulgent policies that reinforce the continued exploitation of the natural world.

Though the dominant thinking of the times does not determine what a novel will say, it does condition content and form, as well as influence how popular a novel will be. Challenging novels like Bambara's *The Salt Eaters* and Walker's *The Temple of My Familiar* seem in significant ways to run against the contemporary American grain. Both posit a complex interrelation between public and private life and invite readers to confront the dilemmas of living in the last quarter of the twentieth century without easy or programmatic solutions. Significantly, neither has received the endorsement of the popular culture that still seeks easy answers to evermore complicated questions about the future of race relations and human relations in the next century.

Clearly, Toni Morrison's rigorous adherence to the historical facts of the African-American experience is key to her portrayal of characters whose consciousnesses are largely informed by the ahistorical imaginative and poetic world of personal experience and inherited narratives. On the other hand, the power of Alice Walker's folk history is diminished somewhat by inaccurate historical detail, such as having Tubman president of Liberia years before he took power in 1944. Similarly, the historical and chronological component of Sherley Anne Williams's *Dessa Rose* is weakened by her confusing the time of Nat Turner's rebellion and even by her placing William Styron's *Confessions of Nat Turner* in the 1970s rather than in 1967.

Although this study focuses on the way these narratives relate to the larger historical context, I do not mean to suggest that a given novel's value as a work of art is necessarily determined by the degree to which it employs historical detail. I do insist, however, that its power is intensified by historical accuracy. While I confess to a preference for fictional narratives that are circumstantially and chronologically correct, my focus on those elements is not intended as a dismissal of the power of the highly personal, subjective, poetic passages of *Dessa Rose* or of the impact of Celie's idiosyncratic voice in *The Color Purple.* My judgments about overall artistic value are largely implied, rather than discussed at length. There is no question in my mind that the novels of Toni Morrison belong in the category of major American fiction and that Shange's *Betsey Brown,* though set in a particular time and place—St. Louis in 1959—is a lesser novel, in part because it lacks the richness of texture created by the integration of substantive historical detail with in-depth development of character. Moreover, while I judge *A Measure of Time* and *A Short Walk* to be substantial literary accomplishments as well as important cultural documents, I suggest that they have not received the attention they deserve by readers or critics because they belong to a genre that is not especially prized by the culture at this time.

Barbara Christian has recently admonished critics "to let go of their distanced and false stance of objectivity and to expose their own point of view."[2] My predilections are for fictional narratives that are grounded in rigorous, historically accurate apprehension of time and place, narratives that consider private lives in the context of public history, which inevitably creates the contingencies shaping those lives.

To undertake a project that focuses on black women writers is to enter a many-voiced discourse in which issues of race, class, and gender are being explored in ever-new combinations. Central to the dialogue is black feminist criticism. Hazel Carby, in her introductory chapter to *Reconstructing Womanhood: The Emergence of the Afro-American Woman Writer* (1987), traces this discussion from Barbara Smith's "Toward a Black Feminist Criticism" (1977) through the qualifications proposed by Deborah McDowell in her 1980 essay "New Directions for Black Feminist Criticism" to Barbara Christian's *Black Feminist Criticism: Perspectives on Black Women Writers* (1985).[3] While presenting the major features of these critics' positions and pointing out what she considers fallacies in each, Carby warns against a transhistorical, essentialist epistemology that she finds in too much black feminist writing and urges a responsible, particularist, historical theory and methodology. Concerned with the historically verifiable conditions in which nineteenth-century black women writers constructed literary texts and in the process reconstructed their own futures by intervening in the "social formations" of their own time, Carby repudiates those critical enterprises that presuppose an essentialist and reductive black female language and literary tradition.[4]

Responding to Carby, Barbara Christian indicates that Carby's view of history is different from her own, which she considers to be less rationalistic, more intuitive and creative, but history nonetheless: "One must ask whether the study of an intellectual tradition necessitates the denial of an imaginative, creative one? Who is to say that the European emphasis on rational intellectual discourse as the measure of a people's history is superior to those traditions that value creativity, expression, paradox in the constructing of their historical process?"[5]

Carby considers attempts to describe a black women's literary tradition to be reductive and usually neglectful of the social and historical contingencies at work in literary texts; others regard the effort to establish a black women's literary tradition a priority for critics. Contributing to the effort to discover how black women's texts relate to each other and to other African-American texts is Henry Louis Gates, Jr. In the last chapter of *The Signifying Monkey*, Gates applies his theory of signifying to the intertextual implications of Alice Walker's *The Color Purple* and Zora Neale Hurston's *Their Eyes Were Watching God.* In the process of examining the connections between the two texts, Gates distinguishes Walker's method of paying homage to an antecedent text from that of Ishmael Reed and others who signify through parody, by implication criticizing other texts by black writers in order to expand their own narrative possibilities.[6]

Conjuring: Black Women, Fiction, and Literary Tradition (1985), edited by Marjorie Pryse and Hortense J. Spillers, contains essays by some fifteen critics concerned with placing individual novelists in a larger, largely literary—and often feminist—context or tradition. Referring to the volume as "a beginning in critical definition," Pryse insists that "there remains almost everything yet to be accomplished" in the effort to integrate the cultural enterprise of black women's studies into the academy. In her afterword to the volume, Spillers points to the common assumptions about the tradition in Claudia Tate's *Black Women Writers at Work*, Barbara Christian's *Black Women Novelists: The Development of a Tradition, 1892–1976*, and Mary Helen Washington's first two anthologies, *Black-Eyed Susans* (1975) and *Midnight Birds* (1980). In suggesting directions for future studies, Spillers alludes to the complexities critics will encounter as they set black women's texts "against the general background of African-American life and thought." She endorses projects that undertake to place this literature in the context of other parallel writing communities as well as those intended to discover "the elaborate and submerged particularities of the texts from writer to writer and within a writer's own career." As Spillers and Pryse saw it in the mid-1980s, the field was wide open.[7]

Indeed that there is a field for black women's literature was far from apparent less than two decades earlier. In 1970, Toni Cade [Bambara] could rightly claim that the publication of *The Black Woman*, an anthology of poems, stories, and essays by black women, was "a beginning";[8] and Mary Helen Washington correctly predicted in 1975 in the first of her ground-breaking anthologies that there would be "quite a revelation in the country of the black woman writer, for the territory is still wilderness."[9] That revelation came with the publication of The Shomburg Library of Nineteenth-Century Black Women Writers under the general editorship of Henry Louis Gates, the Black Women Writers Series published by Beacon Press under the general editorship of Deborah E. McDowell, and various texts emerging from the Feminist Press and some university presses. Hardly anyone could have predicted, however, the abundance of new books by black women that would come forth after Washington's statement—all but five of the novels considered here.[10]

Five years ago, when I began this book, I felt that I was venturing into virgin territory by even raising questions about how narratives reflect, grow out of, or examine the complex set of phenomena that make up the struggle for racial justice. Since then others, including Hazel Carby, have completed projects that confront in different and somewhat larger contexts some of the issues considered here.[11] While I acknowledge that issues of gender, class, feminism, canonization, and literary theory are all relevant to a global comprehension of a body of literature, my aim is limited, linear, and concentrated. The big picture, I believe, will not be a panoramic and synthetic view of carefully integrated parts, but a pastiche of bits and pieces periodically held together in the tension of a moment, only to be rearranged under the pressure of changing times. My intention is to provide one of those pieces by demonstrating how these eighteen novels published since the peak moments of the civil rights movement relate to the movement and to the historical contingencies that fostered it and led to its decline.

CHAPTER ONE

SLAVERY AND RECONSTRUCTION

Three novels by black women published in the past quarter-century are set in the final years of slavery and its immediate aftermath. Margaret Walker's *Jubilee* (1966), though it focuses on slaves and slavery, celebrates the century of progress made by the progeny of slaves—however gradual and however marred by losses and periods of reactionary backlash—extending from the end of the Civil War in 1865 to the end of the movement as it was defined by the passage of the Voting Rights Act in 1965. Sherley Anne Williams's *Dessa Rose* (1986) is a tribute to those who escaped from slavery and went west to settle in a region that, though it was still plagued by racism, was not haunted by the slave past. Although there are survivors in Toni Morrison's *Beloved* (1987) who will presumably become the forebears of the black middle class in the late twentieth century, the emotional power of the novel comes from its memorial for those who did not survive. In each of these novels, the focus of the historical perspective mandates the narrative structure and the treatment of character.

As a celebration of racial progress, Walker's sweeping overview of individuals caught up in widespread and revolutionary events requires the linear progressive narrative associated with much traditional historical fiction. Its characters are representative and typical of their time in history and place in society. Conceived in Walker's childhood when she promised her grandmother that she would someday write her great-grandmother's story, *Jubilee* was in progress for some thirty years. In "How I Wrote Jubilee" (1972), Walker explains that she sees herself as "a novelist in the role of social historian," and confesses that, from the time she promised her grandmother she would tell the story, the commitment to write the book became "a consuming ambition, driving me relentlessly."[1] Ostensibly because she was determined that her story be historically accurate, Walker spent years, even decades, studying Civil War histories and primary documents "to authenticate the story" her grandmother told her.

It probably was not, however, Walker's confidence that she now had sufficient knowledge of the past but rather her sense that this was the appropriate moment for her story to be told which enabled her to complete her novel in 1965 in the midst of the culminating moments of the civil rights movement. Only when she and other African Americans were finally enjoying a kind of victory in the hundred-year-old struggle against racism that had been raging since the end of the Civil War was she able to bring her story to a conclusion that is both celebratory and, in terms of the mid-1960s, historically sound. By leaving her characters planning for the future in the days immediately preceding the widespread establishment of legal Jim Crow practices, Walker concludes with a time and with social conditions that were very similar to those of 1965, when she was writing the final words of this novel that had consumed so much of her life. That year it was possible to look back at slavery and then forward to see the promise of continuing progress, to imagine that the accomplishments of the modern civil rights movement began with the efforts of people like her characters—Vyry, Innis Brown, and Randall Ware—and to celebrate that long struggle with a fictional centennial, a jubilee.

By using a nonlinear narrative structure and shifting narrative voices, Sherley Anne Williams's *Dessa Rose* focuses on individual consciousness and personal experiences, as fugitive slaves repeatedly flee and finally, in 1847, escape their bondage during a particularly repressive period. Some twenty years after the publication of *Jubilee*, at a time when the gains of the civil rights movement were being threatened on all sides and there was less cause for celebration than in the mid-sixties, Williams chronicles the adventures of slaves whose rebellion ends in a kind of personal, though limited, freedom in the West, outside of mainstream society.

Toni Morrison's *Beloved*, on the other hand, remembers those who were totally destroyed or severely crippled by the experience of slavery. Rather than focusing on representative or typical slaves, as does *Jubilee*, or on the extraordinary individuals whose escape from the depths of slavery led almost miraculously to satisfactory private lives, as does *Dessa Rose*, *Be-*

loved examines the highly personal and historically determined lives of slavery's most afflicted victims before and immediately following emancipation. With a linear narrative firmly fixed in 1874–75 as a base, the characters' pasts are woven into the primary story so continually and subtly that the tyranny of that past impinges on the experience of the linear narrative just as the characters have felt it interfere with their lives. Much more than Williams, Morrison has grounded her story in history—the Fugitive Slave Act, the workings of the Underground Railway—and her narrative reveals how the characters' lives are conditioned by historical circumstances. Even though some of the characters are aware of the communal and public contingencies that shape their private lives, and even though in the pre–Civil War days some are active in the Underground Railway, by 1874 not one of the characters is playing a role in public life.

In the afterglow of the civil rights movement, Margaret Walker's text suggests progress and the possibility of effective social action. Williams's novel, appearing in the "Reagan retreat," glorifies those who act heroically regardless of history. Morrison's highly acclaimed work—a eulogy for those who were destroyed by slavery, as well as those who continued after freedom to suffer its consequences—calls attention in the late 1980s to the stories of history's casualties, suggesting that most slaves did not prevail in the aftermath of emancipation, any more than most ghetto dwellers will prevail in modern American society. Speaking of her dismay that there was no "piece of art that commemorates, remembers all . . . the innocent black dead," Morrison offers *Beloved* as such a monument.[2]

Jubilee

Margaret Walker's *Jubilee*, in many ways a conventional historical romance, has an interesting history of its own. When Walker was a small child living in Birmingham, Alabama, her born-in-slavery maternal grandmother told endless stories about her own mother's life before, during, and after the Civil War. To fulfill her promise to tell the story of her own great-grandmother, Walker created three generations of women: Hetta, the black mistress of the plantation owner; their daughter Vyry, the protagonist, modeled after the great-grandmother; and her daughter, Minna, the fictional counterpart of Walker's grandmother, who first told the stories.

Walker first began to write down the stories in 1934, when she was still a college student. Years and then decades passed before she completed the novel. While raising a family, attending graduate school, publishing poetry, and teaching English, Walker periodically immersed herself in the history of plantation life, slavery, the Civil War, and Reconstruction. She read histories, slave narratives, diaries, letters, and other personal documents; studied Civil War newspapers, pamphlets, and songs; and traveled to the places her grandparents lived, delving into records of real estate transfers in the court house in Dawson, Georgia. In the early 1960s, as the "Civil War Centennial approached," she reported feeling "desperate to finish" the novel.[3] She was more than fifty when she fulfilled her promise in 1966 and published *Jubilee.*

In the winter and spring of 1965, during the same months that civil rights activists were marching in the streets of Selma, Alabama, Walker wrote parts 2 and 3—approximately two-thirds of the novel—in a frenzy of creativity and concentration. The events of those climactic moments of the civil rights movement seem to have informed the novel as a whole and the final part in particular. The Voting Rights Act was passed in August 1965, as Walker was preparing *Jubilee* for the publishers. By ending *Jubilee* in 1870, a period of relative calm, she leaves her characters on the brink of events that would lead to decades, and even another century—two more jubilees—of betrayals and violence. She was close to producing a completed draft at the end of March, when Martin Luther King, Jr., stood on the steps of the Alabama state capitol warning that blacks "are still in for a season of suffering," and repeating the words of "The Battle Hymn of the Republic," stressing the song's final lines: "Be jubilant, my feet, our God is marching on . . . His truth is marching on."[4] Even though she worked at breakneck speed that winter and spring, writing from 7:00 A.M. to 11:00 P.M. and pushing herself "beyond all physical endurance," Walker surely noted the front-page news from Selma, Montgomery, and Washington, D.C.[5]

The almost fifty years from the time her grandmother first told those tales until the publication of *Jubilee* correspond to the period when the foundations of the modern civil rights movement were laid: by the National Urban League, formed in 1911 to find ways for blacks to have equal opportunity for housing and employment; by the NAACP, which was founded in 1910 and began organizing in the South in 1917; by the Garvey movement, 1916–23, which glorified racial pride and demonstrated that a mass movement was possible; by the Legal Defense and Education Fund, established in 1939 to finance the court battles that eventually led to *Brown v. Board of Education;* by the Congress of Racial Equality, which staged a number of nonviolent, direct-action protests throughout the forties; by A. Philip Randolph's March on Washington Movement of 1941 and his call in 1943 for mass protests modeled after Gandhi's passive resistance movement in India; and by the lifelong efforts of such individuals as Mary McLeod Bethune, W. E. B. Du Bois, Bayard Rustin, and Ella Baker. During the more than thirty years that Walker was struggling to tell the story of the men and women a century before who longed for freedom and fought to claim it, most of the battles in the war for civil rights were fought.

Jubilee is set in rural southwest Georgia and nearby Alabama communities, the same part of the South associated with the extreme racism that was the target of important campaigns of the civil rights movement. Vyry, the protagonist of the novel, is the child of a white plantation owner and a slave woman; the novel recounts her life from her birth around 1834 to a new beginning at the end of the summer of 1870. For some readers, as for some reviewers, *Jubilee* may seem to be little more than a conventional romance, complete with the cliches of the genre: the elaborate entertainments on the plantation, the patriotic lady who runs the plantation and bravely sends her son and her son-in-law off to die in the war she believes is fought to defend her way of life, emotional deathbed scenes, a love triangle in which a woman must choose between the man she loves and the one to whom she owes her life. Hazel Carby has observed that *Jubilee* was "a particular response to the dominant ideologies of the popular imagination embodied in Mitchell's *Gone with the Wind.*" But Walker has used the elements of popular romance to create a very different kind of novel, one that celebrates not the Old South, but the twentieth-century struggle for civil rights. Her history is not the story of a dead past, but of the past as the precondition of the present. As Carby has observed, "Walker's representation of slavery is her philosophy of history, which is to be understood as the necessary prehistory of contemporary society."[6]

Rather than being haunted by the past, Vyry lives in the present, only occasionally looking back from her bustling kitchens to all that she has lost. She moves from one tragedy—and even one beloved husband—to another with some sadness and difficulty, but she is rarely incapacitated by suffering. A celebration—a jubilee—of what Maya Angelou calls "the heroes and sheroes" of the past, this novel does not belabor the most devastating experiences of slavery. Even the horror of Dutton's sexual exploitation of Vyry's mother, Hetta, is mitigated by his visit to her deathbed to comfort her and by his grief when she dies. Only once, after suffering a miscarriage and losing her house in a fire, does Vyry experience despair.

Unlike even Scarlett O'Hara, she is never hungry. Though the overseer beats her when she runs away, Vyry is neither raped nor abused by the men in her life. She is more like Faulkner's Dilsey than the heroines of other black women novelists—Sophia in *The Color Purple* or Pilate in *Song of Solomon*, for example.

Concentrating on the conventional aspects of the novel, some critics have ignored its relevance to the mid-1960s. Although *Jubilee* does not highlight the ugliest conditions of slavery, it also does not celebrate the Old South but focuses on the slaves who live on or near the plantation owned by John Morris Dutton, a wealthy white planter who has two children by Salina, his socially prominent wife from Savannah, and some fifteen others by Hetta, the black woman his father gave him when she was still a young girl. Of Hetta's children, only Vyry, her mother's favorite, becomes a part of his life; the others he sells or apparently forgets. Events in this novel about slavery and its aftermath, and even the form of the novel, are directly related to the struggle for civil rights that peaked in the mid-1960s.

Each of the novel's three parts could stand alone as a novella. "Sis Hetta's Child: The Ante-Bellum Years" begins by exploring everyday events of slave life as Hetta is dying in childbirth and ends as Vyry recovers from the beating she endured after attempting to run away. "'Mine Eyes Have Seen the Glory': The Civil War Years" focuses on life in "the big house" during the war, relates the destruction of the white family and degeneration of the slave society, and ends with the closing of the plantation and the departure of the slaves. "'Forty Years in the Wilderness': Reconstruction and Reaction" begins in 1866 as Vyry and her children leave the plantation with her new husband Innis Brown in search of land, work, and education for the children and ends in 1870 after they have finally found these and other benefits of freedom.

Even in part 1, "The Ante-Bellum Years," the narrative chronicles progress. Vyry, for example, fares better than her totally oppressed mother. As the mistress of John Dutton, who has used her sexually since she was fifteen years old, and the wife of a black man Dutton has forced her to marry, Vyry's mother, Hetta, does not even have the freedom to decide who uses her body. After some fifteen debilitating pregnancies, she dies in childbirth before her thirtieth birthday. Though Dutton exploits his daughter by bringing her to the big house where she serves first as playmate to her own half sister, then as a kitchen helper, and finally as the family's cook, he does allow her to maintain a relationship with Randall Ware and to raise their children. And he promises to free her—but only after his death.

But progress within slavery is severely limited. To run away with the man she loves, Vyry would have to abandon her children. When Ware begs Vyry to leave with him, her maternal feelings override the desire for freedom. At the end of part 1, after she has made a half-hearted attempt to escape with a small child and a suckling baby, Vyry struggles to regain consciousness after being beaten. She recalls seeing John Dutton standing over her "cursing terrible oaths" (145). This image of a subdued black woman examining her almost mortal wounds and a white man cursing is emblematic of the condition of all subservient people whose freedom is enjoyed at the indulgence of their oppressors. At the end of part 1, Vyry has exhausted the limits of personal freedom within a slave society. Though it is not clear whether Dutton is cursing Vyry for trying to run away or his white workers for beating this woman who is, after all, his own child, the motives for his curse are irrelevant. Within the slave society, Dutton, like all oppressors, will inevitably curse those he oppresses.

Written before the civil rights movement's final push for federal legislation, part 2 belongs to that stage in the long struggle for liberation when white supremacy prevailed. When Ware writes to Vyry that there will be a war to free the slaves, she retorts: "A war to set us niggers free? What kind of crazy talk is that?" (166). At this point Vyry sees what Walker herself perhaps saw when she first conceived of the novel: a world in which white people make promises they never keep and black men join the struggle for freedom, leaving women behind with the children.

In part 2, "The Civil War Years," composed in early 1965, Vyry's desire for freedom is replaced by a struggle for survival, as she works to hold together the white family on which she and her children depend. In the course of the war the Dutton family slowly disintegrates: Marse John dies after an injury; his son and son-in-law receive fatal wounds in battle; with the sound of "big Yankee guns" firing in the distance, Miss Salina suffers a deadly stroke; and her daughter, Lillian, sustains a head injury and permanent brain damage when a Yankee soldier rapes her. As the world collapses around her, Vyry determines "to plant some kind of crop," since she has "the younguns to feed"—Miss Lillian's and her own (231).

Scenes in which characters fail to recognize the relationships between their own private experience and sweeping public events surface here and repeatedly in black women's fiction written since *Jubilee.* The limitation of

Vyry's exclusively private vision is most conspicuous when Union soldiers come to read the Emancipation Proclamation to the slaves. Preoccupied with feeding the crowd and caring for the children, Vyry does not understand the implications of this momentous public event. And at the end of the war, ignorance of the public arena leads Vyry to lose the man she loves: when Randall Ware does not return immediately, Vyry accepts the protection and eventually the love of Innis Brown, a kind but less compelling man.

In the final scene of part 2, on Christmas day of 1865, Vyry prepares two feasts. She serves "baked fresh ham and candied sweet potatoes and buttered whole okra and corn muffins and pecan pie and elderberry wine" to the white folks in the dining room, while she, Innis, and the children eat "possum with sweet potatoes and collard greens and okra and . . . sweet potato pone" in the kitchen (257). Like the tableau at the end of part 1—a white man standing over a subdued black woman—this scene is emblematic of the nature of race relations at a particular time: two families, each without a father, celebrating Christmas in separate rooms with different meals. Both the structure and the imagery of the fiction are determined by history. The dinner is in a sense both the last of an old way of life and the first of a new; from the perspective of the mid-1960s the alienation of the races began when masters and former slaves first chose to sit down at separate and not-so-equal tables.

In part 3, written during the climactic moments of the southern civil rights movement, public attitudes and events increasingly impinge on the lives of the characters as whites engage in violence to limit the freedom of blacks, who in turn struggle to claim the rights promised by freedom—housing, food, employment, education, and involvement in the political process. Having won the battle of "freedom from," they now must fight the battle of "freedom to." When Vyry and Innis set out with her two children, Jim and Minna, their plight is the same as that of "hundreds of thousands of emancipated Negroes" (263). Locating what seems like an ideal spot for farming, Vyry and Innis build a cabin, plant crops, and enjoy a plentiful harvest. But the following spring the river floods their house and fields, and they soon are on the road looking for a new place to settle. In the next three years, Vyry and Innis are cheated, exploited, and terrorized by poor whites who are determined to prevent blacks from owning farms and competing with them for jobs. Light enough to pass for white, Vyry contributes to the family income by selling eggs to white women who do not know she is a mulatto. After Vyry assists one of these women in childbirth, her fortune changes. In need of a midwife, the white families urge Vyry to stay in their community, and the men volunteer to help Innis build a house.

The day of the house-raising is idyllic. The women bring quilts; the men work all day building the new house. Just as she had done that last Christmas on the plantation, Vyry prepares a feast: peaches, dew berries, sweet cream, ham, eggs, fried chicken, biscuits, buttered corn, greens, okra, blackberry pie, and coffee. In striking contrast to that earlier segregated meal, which served as a symbolic vignette at the end of 1865, blacks and whites now sit together and enjoy the food and fellowship. The first half of the summer of 1870 is "full of halcyon days one dreams and scarcely believes are real" (371). But before long the stresses of unrelenting economic pressures and unending labor begin to take their toll. Innis, frustrated by Jim's laziness, lashes out in anger and beats him brutally.

Randall Ware, meanwhile, very much alive, has delayed his return to private life to attend the First Convention of Colored People in Georgia. Putting his public responsibilities before his private ones, he joins the Georgia Equal Rights Association and determines to take "an active part in the political affairs of his county, town, and state" (270). It is during Ware's delay that Vyry despairs and leaves with Innis. Just as Vyry's exclusively private vision robs her of the power of public action, Ware's public commitments result in the loss of his private world. Had Vyry known about the scramble for political power that followed the war and had she understood Ware's commitment first to freedom and then to power, she might have concluded that he was involved in the struggle and would take care of his private life in due time. On the other hand, if Ware, out of consideration for Vyry's personal needs, had gotten word to her, she would surely have waited for him. But in fact neither finds a way to balance public and private responsibilities.

At first Randall Ware, like Vyry, underestimates the tenacity of racism. As a propertied free black, Ware is a ready target for racist terrorism. When he returns to claim his land and forge, the whites threaten him and demand that he sell them his land. When he refuses, "white-sheeted callers" arrive at his house, throw the bloody body of his journeyman at his feet, beat him, and leave him half-conscious in the woods (327). Defeated and frightened, Ware abandons his plans to play a part in the public arena, and he recognizes that a new war has only just begun: "This is a war of white against black and it's a night war with disguise and closed doors. The first white man you see in the morning could be the very man who beat you within an inch of your life the night before. No, they have begun a reign of terror to put the Negro back in slavery. They will never accept the fact that the South rose up in rebellion against the Union North and the North won the war. They mean to take out all their grudges on us" (333). In certain communities—Selma, Alabama, for example—these words were as appropriate in the mid-1960s as they were a hundred years before. But unlike many of his counterparts in 1965, Ware sees no way for direct political action.

Toward the end of the novel, Ware, hoping that Vyry will come back to him and his son Jim will go away to school, searches out his family. Surprised and shaken by Ware's visit, Vyry soothes her uneasiness by preparing a welcoming feast. Once again, a dinner-table scene is emblematic of the condition of a community, this time the extended black community. At the end of the war Vyry prepared for her two families two meals served at separate, segregated tables; during that brief period after the war when some whites and blacks recognized thier mutual dependency, she enjoyed one jointly prepared, integrated meal; now, members of her family, representative of the larger black community, sit down at the table together. Their differences are subsumed in a larger commitment to each other's welfare.

Randall explains that he had run for the state legislature and won but that "white folks couldn't and wouldn't stand for it" (400). Having abandoned his dream of playing a role in the public arena, Ware adopts a different plan for bringing about change. The first step, he argues, is education: "And so far as education is concerned, I tell you it may not be the only way for our people but it is the main way. We have got to be educated before we know our rights and how to fight for them" (404). By putting education—a top priority for many contingents of the modern civil rights movement—as the first item on Ware's agenda, and by not having him advocate retaliatory violence, Walker establishes him as a predecessor to some contemporary nonviolent movement leaders; much of his rhetoric, however, is consistent with that of black militants of the mid-1960s. Like many movement activists Ware is uncertain about the best means to achieve his goals.

The differing positions of the characters in this final part of the novel are almost parallel to those of contemporary reformers as Randall, Innis, and Vyry each voice one of three conflicting positions dividing the black community. Randall Ware's views, though sometimes moderate, are more consistent with those of the emerging separatists who by 1966 would take over the leadership of SNCC. Ware argues that blacks will have "to fight and struggle" for "education, land, and the ballot" (396), that the "average white man hates a Negro, always did, and always will," and that "every white man" believes that every black is inferior and should be treated like "a brute animal" (397). Like Malcolm X, he never advocates specific acts of violence, but he does use militant language, equating the struggle to come with the "years of fighting and struggling" that were necessary to end slavery (396). Like SNCC leaders in 1965 who were arguing for the expulsion of whites from the organization, Ware opposes any cooperation with whites and urges a kind of 1870s black power.

Innis Brown, on the other hand, argues that blacks must accommodate to life in a racist society by relinquishing any hopes of equality: "They was a man not so long ago made a speech round here and he says the colored peoples got to forgit about the political vote and tend our farms and raise our families and show the white folks we ain't lazy and ain't stirring up no trouble for nobody, but we is for peace and we's good citizens. . . . I kinda believes like that man" (399).

As a mulatto, Vyry, on the other hand, represents that integrated society dreamed of by Martin Luther King, Jr.[7] She insists that all people are capable of good and that being "apart and separated from each other" makes people hate (397). As she sees it, the solution to the racial problem lies in blacks and whites acknowledging their interdependency: "They ain't needing me no worser than I is needing them, that's what. We both needs each other. White folks needs what black folks got just as much as black folks needs what white folks is got, and we's all got to stay here mongst each other and git along, that's what" (402). Each position might be summed up in a single word: separatism, accommodation, and coalition. The novel itself, however, does not validate one view over the other but rather gives each a voice, suggesting that any solution to the problems of racism, if there is to be one, whether in 1870 or 1966, will inevitably evolve from the clash and resolution of such differences and the conflicting assumptions on which they are based. As the protagonist of the novel, however, Vyry stands between two extremes, and her advocacy of integration and interdependency, compatible with the position of Martin Luther King, Jr., seems to have the greatest weight.[8]

As Vyry, Innis, and Randall sit talking through the night, they all have their say, and the conflicts that divide them, though unresolved, are set aside for the sake of communal goodwill. Walker completed *Jubilee* before the Watts riots in 1965, but her first readers, still reeling from the violence that erupted in August of that year, may have had difficulty imagining how such apparently mutually exclusive positions would ever be reconciled in a jubilee of good feeling.

Jubilee is mostly a story of the 1860s for the 1960s. In 1966, the year that *Jubilee* was published, blacks had begun to enjoy full access to public accommodations, to go to the polls occasionally to elect black officials, and to see the racial barriers fall in public schools and in colleges. The Civil Rights Acts of 1964 and 1965 had given legal sanction to the gains of the struggle, and for a while it seemed that progress would continue. Vyry's condition at the end of *Jubilee* is similar to that of her counterparts a hundred years later. She is not only free from slavery, but she has learned to cope with the everyday responsibilities and to enjoy limited but significant privileges of freedom: a home, work, a loving husband, and an education for her children.

The relationship between the post–Civil War period and the modern civil rights movement emerges in the final pages of the novel as this family—Vyry, her two husbands, and the children—becomes virtually allegorical, with the three adults each representing a major faction of the black community and the children representing the future. Soon after the war, former slaves were systematically deprived of their civil rights through a series of compromises, legislative acts, and court decisions, culminating in the validation of the separate-but-equal doctrine of *Plessy v. Ferguson* in 1896. While the final events leading to the restoration of these rights were occurring, as Walker completed her novel in 1965, the language and place names associated with those events came forth in the text, connecting the present with the past. Vyry talks of being "free at last" (405), and Ware of having had "a dream" (409). Jim, on his way to Montgomery and then to school in Selma, Alabama, is literally and figuratively beginning a journey that will pass through Montgomery and Selma and culminate in the civil rights legislation of 1964 and 1965. When Jim boards the train—its first stop Montgomery—and the "white trainman" shouts, "Colored up front! White ladies to the rear," he is experiencing the early days of the Jim Crow system that will remain intact for nearly a century (415). The first major battle of the movement was fought over the issue of seating on the city buses of Montgomery, the very city to which Jim is heading.

The final scene of the novel looks both forward and backward. Vyry, simultaneously grieving over her son's departure and waiting expectantly for the birth of another child, goes out into the dying light and looks "over the red-clay hills of her new home" (416). She recalls herself many years before as a little slave child chasing her master's dominicker hen while Aunt Sally calls her back to work in her white master's house, and for a moment she feels "peace in her heart," as her own "flock of white leghorn laying-hens come running" when she calls (416).

The jubilee of Old Testament law mandates the release of slaves from bondage every fifty years. Like history, it suggests caution, for the battle for freedom will have to be fought again and again. Shortly after the close of the Civil War, plans were already underway to reinstitute slavery in the more subtle form of Jim Crow. The next "jubilee," fifty years after the Civil War, came roughly at the time of the First World War, when those African Americans who went to France and survived the war enjoyed some freedom from Jim Crow and came to expect freedom from racism at home. Lynchings and other acts of terror tightened the shackles again as soon as the Great War ended. Just before *Jubilee* was published, some fifty years later, the promised jubilee came forth in the culminating events of the movement. The title, then, may be both a celebration and a warning. While linking the contemporary movement to the past, the novel anticipates the need to adjust to new freedom and to prepare for a backlash from threatened whites determined to re-establish the conditions that existed before the civil rights movement. Readers in the 1990s are more likely than their 1960s counterparts to recognize the implied warnings and to acknowledge the accuracy of its prophecy.[9]

It seems likely that Margaret Walker, who by her own admission had been writing this book all her life, was able to draw on energy generated by the urgency of the movement itself to bring it to fruition, but perhaps in those heady days she was able to end the novel with a more positive commitment to progress than would have been possible for her later. As the great-granddaughter of a former slave, Walker's own life has reflected the progress made possible by those women in the nineteenth century who struggled to create the homes and families that would nurture growth and development. In the spring of 1965, while blacks and whites were confronting each other on the Edmond Pettus bridge in Selma, Klansmen were murdering civil rights workers of both races on the back roads of the South, and blacks themselves were gearing up for an internecine struggle, Walker was completing with breakneck speed this book that celebrated progress while calling for reconciliation, cooperation, and commitment to common goals. Before he leaves Vyry at the end of the novel, regretting the loss of his wife and children, Randall Ware observes that they "just got caught in the times" (409). In 1965, James Reeb, Jimmie Lee Jackson, and

Viola Liuzzo—all killed during the Selma campaign—were caught in the same stream of history.

Some readers in 1966 were likely to have experienced the novel as a celebration of the linear progress toward racial justice; a generation later, we know that regression is as likely as further progress. Margaret Walker once promised a sequel to *Jubilee*. One wonders what its vision of the state of racial justice would be today, or what future it might project.

Dessa Rose

In the mid-1960s Margaret Walker could look back at the Civil War and Emancipation as progressive steps in the political and social movement from slavery toward freedom, and perhaps even a nonracist, integrated society. The movement of the 1950s and 1960s was a later, further step in the onward march to that end. Twenty years later in the 1980s, to Sherley Anne Williams, some thirty years younger than Margaret Walker, the movement that was vital in the mid-1960s seemed to be spent and, if honored, no longer relevant or useful. Social action—history—seemed to end in a swamp, not a river rushing inexorably to the sea. Only the heroic or extraordinary acts of the individual could make life better for beleaguered African Americans.

Williams's *Dessa Rose*, however, like *Jubilee*, is set in the South in the mid-nineteenth century and is based on historically documented events. Like *Jubilee*, it was a long time coming to fruition. In the early 1970s Williams wrote a short story set in northern Alabama entitled "Meditations on History" about a slave girl who attacked her master, participated in a violent rebellion, and twice escaped from her shackles. This story did not appear in print until 1980, when Mary Helen Washington included it in *Midnight Birds*. Though "Meditations" is the basis for "The Darkey," the first of the three parts of *Dessa Rose*, it lacks the vitality and lyric momentum of the longer narrative. Focusing entirely on the interaction between a rebellious slave girl named Dessa and Nehemiah, a white male historian who is interviewing Dessa for his book on the causes and ways to eradicate slave rebellions, Williams might have entitled her story "Meditations on Rebellion." Alternating between Dessa's musings and Nehemiah's naive interpretations of her tales, the story is about how Dessa attacks a white man who has killed Kaine, the young slave whose baby she is carrying. As a consequence, the slaveholder brands Dessa and sells her. While traveling to market in a large coffle of slaves, Dessa murders a white guard and escapes. Later, when she is captured, her life is spared so that she can bear her baby into slavery. While she waits, Nehemiah visits her and listens to her monologues about her love for Kaine, the murder, and the rebellion. Dessa escapes a second time, and the story ends inconclusively. "For myself," Nehemiah says, "I have searched, hunted, called and am now exhausted. She is gone. . . . And I not even aware, not even suspecting, just—gone."[10]

Though some passages in the story and novel are identical, there are significant differences between the two works. "Meditations on History," prefaced by an epigraph from Angela Davis, is set in 1829, the year that the actual slave rebellion took place; *Dessa Rose*, with epigraphs from Frederick Douglass and Sojourner Truth, is ahistorically set in 1847. The story consists of long, meditative passages, while the novel is structured from shorter pieces that jump with Faulknerian speed from one consciousness to another. In the novel Dessa attacks the mistress rather than the master; names are changed, episodes expanded, and characters added. Though Nehemiah plays the same role in both works, "Meditations" ends with him resigned to Dessa's escape, and "The Darkey" with his vow that "the slut will not escape me" (71). Finally, the novel merges Dessa's story with that of a white woman, who, like the pregnant rebel slave woman on whom Dessa's character is modeled, is based on a real life woman mentioned in Aptheker's *American Negro Slave Revolts.* In reality this white woman bore no relation to the woman who is the inspiration for Dessa.[11]

The 1820s were a time when slaves and free blacks repeatedly engaged in violent insurrections, which were usually quelled. In 1822, Denmark Vesey, like Gabriel Prosser in 1800, was hanged without forwarding the cause of liberty for blacks. Nevertheless, in 1829, when the Dessa story was originally set, David Walker, a free black, published his *Appeal*, which called for slaves to revolt and to resist oppression in every way, including armed struggle. Two years later, Nat Turner led another bloody and fruitless rebellion. But these were only the most celebrated of the countless slave uprisings that occurred during the first four decades of the nineteenth century.[12] In the context of such an atmosphere of rebellion, Nehemiah's proposed work on the roots of rebellion may well have been the best-seller his fictional publisher anticipates.

Written in the early 1970s, the short story "Meditations on History" is understandable as a response to the acts of violence that marked the period following the civil rights movement. Like the slave rebellions of the 1820s and 1830s, the urban riots that came in the wake of the Voting Rights Act in 1965 accomplished little for those involved; and the militant advocates of violence in the sixties, like the leaders of slave insurrections, very often lost far more than they gained. Like many contemporary black militants and like Angela Davis, to whom the story is "respectfully, affectionately dedicated," the Dessa of "Meditations" went into the unknown, a fugitive fleeing her oppressors.

In the novel, however, Williams moves her story forward from 1829 to 1847, from a period of militancy and unproductive violence like the late 1960s to a time when social action is ending and repressive reaction is setting in—a time like the 1980s. By 1847, resistance to slavery had moved from random, unrelated acts of violence to organized efforts to work for long-term change. The first national black convention had met in Philadelphia in 1830; the American Anti-Slavery Society was organized in 1833; the first antislavery political party, the Liberty Party, was established in 1839; Sojourner Truth had left New York and begun her career as an antislavery activist in 1843; and Frederick Douglass had published the first of his three autobiographies in 1845 and launched his journal *The North Star*.

Perhaps Williams changed the temporal setting of her novel because she knew that these public gains in the 1830s and 1840s were met with the Fugitive Slave Act of 1850, the establishment of the separate-but-equal precedent in a school integration suit in Boston in 1850, and the Dred Scott decision in 1857, which in denying Scott his citizenship strengthened the proslavery position. Since by the late 1840s, as in the late 1980s, the organization and mobilization for struggle had engendered powerful reactionary forces, neither aggressive acts nor peaceful public activism seemed to hold much promise for furthering the cause of social justice. In northern Alabama in 1847, whites did not expect abolition and blacks had little hope of emancipation. Improvement in race relations was likely to happen, if at all, only in the personal sphere.

Moving Dessa's story forward from 1829 to 1847, then, changed the context to a time more analogous to that of its composition. The heyday of the black militants of the 1960s was almost exactly the same distance from 1986, the date *Dessa* was published, as the rebellions of Walker and Turner were from the events of the novel. For readers in the 1980s, especially those aware of the history of the 1840s, the temporal setting was analogous to their own undramatic time.

When *Jubilee* was completed in April 1965 it was still possible to write a progressive narrative reflecting the widespread belief that the achievements of the civil rights movement would be the stepping stones to the

Great Society. No one could foresee the devastating effects of the Vietnam War, the repeated assassinations, and urban riots. By the mid-1980s, however, the federal commitment to civil rights legislation had ended. To have written a sequential narrative that suggested continual progress would have been to falsify the current feel and facts of history. In contrast, Williams used the raw material of history in a subjective, nonlinear way.

Dessa Rose is a more modern and modernist novel than *Jubilee*. Focusing on the autonomous self as the neoromanticism of the modernist movement does, it rejects the positivistic realism of the classical historical novel. Its nonlinear structure shifts back and forth in time. The prologue opens with a stream-of-consciousness rendering of Dessa's dreams of loving Kaine. In part 1, "The Darkey," the narrative shifts to the cellar where she is imprisoned and then to narrative voices, including excerpts from Nehemiah's journals and the stories he relates of previous slave insurrections "some seven years before" (27). Part 2, "The Wench," primarily third-person narration, includes passages recording Dessa's dreams of childhood and of Kaine. Part 3, "The Negress," is entirely a first-person narration from Dessa's point of view. In the epilogue, Dessa, now a grandmother, looks back at the days when she and Ruth, her white companion, were bonded friends, working together for common goals.

In spite of its setting in a particular time and place, dates in this sometimes-elusive narrative do not finally matter. Ruminating on Dessa's having participated in a rebellion in which slaves killed white men, Nehemiah observes that he cannot recall a violent rebellion "since Nat Turner's gang almost thirty years before" (29). Since Turner's rebellion was actually only sixteen years before Dessa's, what is significant is not so much the specific period of time, but that Williams creates temporal distance between the time of the novel and the time when slave rebellions seemed a possible way to achieve freedom. By 1847, the methods of Denmark Vesey, David Walker, and Nat Turner had proven ineffective. Though there were significant numbers of isolated uprisings, they usually resulted in death for the rebels.

Outside the narrative, Williams also approximates dates. In the "Author's Note," she refers indirectly to William Styron's *Confessions of Nat Turner* as a work of the early 1970s, when in fact it was published in 1967. Williams's casualness about historical dates de-historicizes, even desocializes the novel, shining the spotlight on the inner and outer lives of individual characters, reinforcing the narrative's focus on private, subjective experience. The modernist *Dessa Rose*, then, is consistent with the conservative political climate of the 1980s, a time when many cultural forces glorified changes within individuals that allowed them to make better lives for themselves, rather than changes in society that promoted change in individuals.

By exploring the possibilities for change inherent in friendship between two women, this highly subjective novel rewrites history into something like myth, the story of slaves and a slave-owning woman who worked together and helped each other. "Meditations on History" ended inconclusively with Dessa's escape and disappearance. The second part of the later novel relates a story very different from that of the slave woman in Aptheker who served as the inspiration of the story. Rather than being hanged, Dessa repeatedly eludes her captors and miraculously finds sanctuary with a white woman willing to help her escape west into a free state. In the "Author's Note" at the beginning of the novel, Williams explains that the stories of the pregnant rebel slave woman in Kentucky and the white woman in North Carolina who aided runaways, recorded in Aptheker's *American Negro Slave Revolts*, inspired her to consider what *might* have been had the two women's lives converged. In the preface to the novel, Williams admits that it may be "only a metaphor" (6). The story, then, may be about the nature of the timeless, ahistorical problems and possibilities of personal relationships beyond race.

The two women's lives converge when Dessa's renegade male companions rescue her from slavery and bring her to the house of Ruth Elizabeth Carson, the daughter of a prosperous estate manager from Charleston and wife of FitzAlbert Sutton. Estranged from her family and abandoned by her husband, Ruth, nicknamed Rufel, lives in isolation with her two small children and her black servants, some of them runaway slaves that she harbors in exchange for their labor. The slowly developing relationship between Dessa and Rufel becomes a metaphor for the integration of society on a personal level. This interpersonal relationship, and not history or historical contingencies, lies at the center of the novel. *Dessa Rose*, then, is about the civil rights movement only in reflecting the shift in the culture from political activism to a concern for personal change.

At the beginning of part 2, when Dessa, exhausted and recovering from childbirth, awakes and finds herself in Rufel's house, she has been dreaming of her child's father and her life back on the plantation. Rufel is nearby alternately suckling her own baby and Dessa's newborn. This reversal of roles—white Rufel playing nurse to Dessa and "mammy" to her black baby—begins the bonding process that culminates in the final scenes of the novel. A serious rift occurs, however, when Rufel is reminiscing about Dorcas, her own "mammy," who has just died. Angry that the white woman presumes to use the name Dessa calls her own mother, Dessa insists that Rufel does not have a mammy. While Dessa forces her to recognize that she does not even know whether Dorcas had children of her own, Rufel still longs to know for sure that Dorcas loved her. When Nathan, one of the runaways, tells her that Dorcas talked mainly about her to the other slaves, Rufel, finally convinced, announces to Dessa that she was loved: "Your mammy birthed you and mines, mines just helped to raise me. But she loved me . . . just like yours loved you" (154). Dessa's response, "I know that, Mis'ess," settles the matter for good (155).

The growing relationship between the two women is threatened again when Rufel takes Nathan, Dessa's friend, to bed. While Dessa claims to be upset because Rufel is risking Nathan's life by taking him as a lover, she also acts like a jealous woman. She sulks and refuses to go along with the plan the others have to escape to free states in the West, insisting that blacks can never trust white people. The others point out that Rufel has treated them fairly: she provides them with a place to live, farms on shares with them, and agrees to a scheme that will bring them all much-needed money. Only after Harker, whose love is like "thunder and lightning" (192), begins to pay attention to Dessa does she let go of her resentment of Rufel and Nathan and agree to go along with the scam. Even the plan to escape to freedom, then, is first endangered and then fostered by intimate, personal relationships.

Part 3, "The Negress," relates the exploits of Rufel and her band of runaway slaves as they travel from one slave market to another. On the road, Nathan serves as driver and Dessa as Rufel's personal maid. Posing as their owner, Rufel sells four of the runaways, who then escape and rejoin the others, only to be sold again and again. Their plan is to make enough money for Rufel to live independently and the others to escape.

The two women finally overcome the obstacles that divide them, and for a brief period before they part forever, they behave like trusting friends. Sharing a room with her mistress on the road, Dessa foils a white man's attempt to rape Rufel and is surprised to discover that white women are also victims of male aggression. Toward the end of the novel, Rufel tells Dessa that she wants to go west with her and the other slaves. Dessa assumes that Rufel wants to be with Nathan, and once again she acts like a jealous woman and lashes out that it's a scandal for a white woman to go "chasing all round the country after some red-eyed negro" (218). It is never clear whether Dessa would be outraged by any white woman who takes a black lover, or whether she is truly concerned that if Rufel comes along with Nathan they would be in danger even in a free state. After Dessa's outburst, she leaves the room and hears Rufel shout, "I'm talking friends" (218).

Out on the street, still trying to sort out whether she and Rufel could be friends, Dessa thinks to herself: "I wanted to believe I'd heard the white woman ask me to friend with her" (219). At that moment, she is accosted by Nehemiah, who has her arrested as a runaway. The incident provides Rufel with an opportunity to prove her declaration of friendship. She not only comes to Dessa's rescue and boldly lies to the sheriff to save her, but afterward, she insists on the equality that is necessary for friendship. Abandoning her childhood name and insisting that Dessa call her Ruth, she almost says to her black friends, "Thy people shall be my people." In order to insure their safety, she decides to accompany them to the western frontier. Though she has concluded that she does not want to "live round slavery no more" (218), Ruth does not follow them into the alien corn. And so she turns back at the border, not to Charleston but to "Philly-me-York—some city didn't allow no slaves" (236).

In the end, the decision for Ruth not to follow the others beyond the border is not hers. In the epilogue, Dessa explains that "Ruth would've tried it" and that, if Nathan had asked her, she would have married him. It was the runaways who decided to go on without her and make their own way. But once they make it to the West, they discover that, even in free states, blacks "can't live in peace under protection of law" without "some white person to stand protection" for them. Though Dessa misses Ruth "in and out of trouble," she consoles herself with the idea that it is not possible to "friend with, love with" someone you depend on for protection (236). The emotional force of this passage is carried by Dessa's grief for a lost, personal friendship.

The final lines of the book, however, for all its person-to-person emphasis, revert to history. Dessa explains in her own words that she has had her story "wrote down" so that she will live in the minds of her children, who must know how they got where they are: "I hope they never have to pay what it cost us to own ourselfs. Mother, brother, sister, husband, friends . . . my own girlhood all I ever had was the membrance of a daddy's smile. Oh, we have paid for our children's place in the world again, and again" (236). Dessa's final words, like the title of the short story that was its genesis, invite readers of the novel to meditate on history, to consider the price that has repeatedly been paid to insure what freedoms her children enjoy. Because the ending is not substantiated by an acknowledgment of the actual historical process by which Dessa's progeny got where they are, the novel seems to suggest that an escape west was sufficient for some blacks to insure freedom for their children and grandchildren.

The struggle then seems to be up to the African-American community, since the European Americans in 1847 and 1986, represented by Ruth, have brought their black friends to the border of freedom and gone back "east," to a safe place where slavery and racism are not their issue. There is little in the novel to invite readers to consider the possibility of public action. Rather, readers are encouraged to identify with Harker, who "feels bad for all them that didn't make it" (149) and to hope with Dessa that their relatively privileged children "will never have to pay" such a price for freedom. Readers may conclude that, in times like these when rectifying the evils of racism has long dropped from the public agenda, there is little to do other than pursue the kinds of freedom that come with personal success.

Beloved

Toni Morrison's *Beloved*, published a year after *Dessa Rose*, explores the complex interrelation between private lives and public circumstances by creating characters whose lives are so intertwined with the exigencies of history that the most private acts and thoughts derive from public policy. Unlike *Jubilee*, neither *Beloved* nor *Dessa Rose* progresses steadily forward in time. Both novels use a complex narrative that moves from the present to the past and back again, sometimes relating events through interior monologues, sometimes in straightforward third-person narrative. The two novels differ considerably, however, in their treatment of the relationship between the personal domain and historically determined events.

Both *Dessa Rose* and *Beloved* grew from stories recorded in the writings of Herbert Aptheker.[13] The protagonist in both novels is a rebellious slave girl who runs away from slavery, bearing the scars of brutal beatings. Both endure the humiliation of being "studied" by white males—Dessa by Nehemiah, Sethe by schoolteacher. Pregnant at the time of escape, each gives birth on the road. Both Dessa and Sethe owe their freedom at least in part to a sympathetic white woman. For Morrison, public policy ultimately determines the nature of personal lives. Characters in *Dessa* have little consciousness of public events. Some in *Beloved* are preoccupied, at least for a time, with events in the greater world. In her first days of freedom, Sethe recalls people waiting for the arrival of Frederick Douglass's *North Star* and hearing talk of "the Fugitive Bill . . . Dred Scott . . . Sojourner's high-wheeled buggy . . . and other weighty issues" (173).

Flashbacks in *Dessa Rose* focus on the emotional and the subjective aspects of the characters' lives, often evoking in lyrical language single isolated vignettes. In *Beloved*, on the other hand, past episodes are continually reinterpreted in terms of the present. While *Dessa* moves to the past then abruptly back to the present, the narrative texture of *Beloved* is more like an intricate weaving with threads from the past being picked up again and again as Morrison presents an episode, incident, or tableau and then explores all that led up to it. *Beloved* is interpretively historic, and *Dessa* mythically timeless.

The climactic episode of Toni Morrison's *Beloved*, reminiscent of the final scene of part 1 of *Jubilee*—Dutton standing over Vyry cursing—ends with a highly visual tableau: a white man rising with a whip in his hand as a black woman raises her hand to kill him. Intervening in this potential violence is a crowd of black women, who some eighteen years before contributed to the woman's decision to commit a violent act so terrible that the memory of it has haunted her ever since. This present intervention is the action which breaks the cycle of violence that has been directing the characters' lives at least since Sethe's escape from slavery.

The novel opens quietly, with an omniscient narrator introducing Sethe and her eighteen-year-old daughter Denver, who live alone in a house haunted by the ghost of a baby. The next few paragraphs fill in the rough outline of events that have led to a summer afternoon in 1873, as Sethe remembers her dying baby's blood pulsating into her hands from its gaping throat; the sexual encounter she bartered for the engraving of the word *Beloved* on her baby's tombstone; her two sons who ran away to escape the baby's ghost; the death nine years before of her heroic mother-in-law Baby Suggs; and the uninvited pictures of the plantation called Sweet Home with boys hanging from "the most beautiful sycamores in the world" that come "rolling, rolling, rolling out before her eyes" (6).

From the reticence of the novel's opening paragraphs to the ambiguities of its conclusion, readers are required to participate actively in integrating the visual images of the characters' memories into the gradually unfolding narrative. For example, the image of boys hanging from trees and Sethe's guilt about remembering the beauty of the "soughing" trees rather than remembering the boys themselves leads readers to "see" the hidden pictures of a lynching in Sethe's carefully edited memories. Through these edited memories and an intermittently helpful narrator, we also gradually "see" the interrelation of events that lead up to and explain Sethe's raising that ice pick with the intention of killing a white man who had once saved her from hanging.

The major features of Sethe's past are in the background of the narrative that begins in the opening pages: As a young slave girl, Sethe "marries" Halle, a fellow slave. When she is pregnant with their fourth child, they join other slaves on the plantation in a scheme to escape. While Halle is captured, Sethe, after enduring a terrible beating and degrading treatment, makes it to the other side of the Ohio River with the help of a white girl who delivers her baby in the bottom of a boat. Safe at last in the home of her mother-in-law, Baby Suggs, a Harriet Tubman figure who aids runaways and inspires fugitives to endure, Sethe enjoys less than a month of freedom and relative happiness before slave catchers come for her. Rather than have her children returned to slavery, Sethe, intending to kill all of them, cuts the throat of her "crawling-already" baby girl (99). The ghost of this child haunts Sethe's house, scaring off neighbors, isolating her from the community.

Sethe's story is only one of several that make up this complex narrative. During a long life as a slave, Baby Suggs had eight children, four of whom were taken from her, four others "chased" (5). Her son Halle worked for "five years of Sundays" to buy his mother's freedom (11), and from 1850 to 1855, Baby Suggs was a leader among her people living in Ohio. Arriving in Cincinnati around 1850, the year of the Fugitive Slave Act, she spends her first five years of freedom preaching, working for the abolition of slavery, and helping runaway slaves—until the community betrays her. After Sethe murders her baby, Baby Suggs abandons her life as an activist and takes to her bed.

The community of friends and neighbors, envious of Baby Suggs's prosperity, collaborate in the violence by failing to warn her or Sethe that slave catchers are in town. Stamp Paid, a former slave who has spent much of his life helping fugitives, remembers Baby Suggs as "the mountain to his sky" (170). But after Sethe's "rough response to the Fugitive Bill" (171), Stamp denounces Baby Suggs for abandoning the struggle. At the end of the novel some of these same neighbors prevent Sethe from killing a white man and empower her to take Baby Suggs's advice and "*lay it all down, sword and shield*" (173).

The opening scene is set eighteen years after Sethe killed her baby. Paul D, one of the "Sweet Home men" whom Sethe has not seen since she escaped, appears on her front porch. Within hours, he has expelled the baby ghost from the house and made love to Sethe. After three days, he insists that they "can make a life" and offers to take care of Sethe and her daughter Denver (46). But that possibility begins to fade with the arrival of a beautiful young woman named Beloved, who saps Sethe of energy and health and bewitches and then seduces Paul D. Beloved, both as baby ghost and as an adult woman, seems to be punishing Sethe for her deed. Stamp Paid, also determined that Sethe must pay, reveals to the unknowing Paul D that Sethe murdered her baby.

The primary narrative of *Beloved* is the love story of a man and woman who meet after eighteen years apart, fall in love, struggle to come to terms with their past lives, and finally see the possibility of making a life together. Woven into this story are tales from the past: of Sethe's life as a slave, her escape, and her encounter with the white girl who delivered her baby; of Paul D's ordeal on a chain gang, his days of agony with a bit in his mouth, and his years of wandering after breaking free. For Sethe the images of the past—the man who made a study of blacks and recorded their animal characteristics; the brutal beating she received when she first attempted to escape; the white boys who, for a prank, constrained her and sucked her lactating breasts; the bleeding child she held in her arms after cutting its throat with a saw—are so powerful that they dominate her consciousness. She cannot imagine ever being free, even for a moment, of what she calls "a thought picture" (36). Listening to her mother's talk about this phenomenon, which Sethe calls "rememory," Denver concludes that "nothing ever dies" and Sethe assures her that "nothing ever does" (36). Most of these "undying" past events emerge from memory—not in narrated, dramatized flashbacks—as characters tell each other their stories. While Sethe fails to tell about killing her own baby, she does tell Paul D much about her past; and he fills Sethe's mind with "old rememories that broke her heart" (95).

When Paul D first asks her to make a life with him, "her brain was not interested in the future. Loaded with the past and hungry for more, it left her no room to imagine, let alone plan for, the next day" (70). Fixing on the stories he tells of his own life as a slave, she sits by him, rubbing his knee: "Like kneading bread in the half-light of the restaurant kitchen. Before the cook arrived when she stood in a space no wider than a bench is long, back behind and to the left of the milk cans. Working dough. Working, working dough. Nothing better than that to start the day's serious work of beating back the past" (73). The horrifying stories that Paul D tells—about Sethe's husband, Halle, being driven mad, about his friend Sixo laughing while he burns to death, and about his own experiences as a prisoner with a bit in his mouth—make the work of beating back the past all the more difficult, but she knows she must if she is to go on with her life: "If she could just manage the news Paul D brought and the news he kept to himself. Just manage it. Not break, fall or cry each time a hateful picture drifted in front of her face. Not develop some permanent craziness. . . . All she wanted was to go on" (97).

But even as Paul D is telling Sethe stories that she must beat back from consciousness, he imagines that he has sealed those very stories away in a safe place: "It was some time before he could put Alfred, Georgia, Sixo, schoolteacher, Halle, his brothers, Sethe, Mister, the taste of iron, the sight of butter, the smell of hickory, notebook paper, one by one, into the tobacco tin lodged in his chest. By the time he got to 124 [Bluestone Road, where he finds Sethe] nothing in this world could pry it open" (113). But living in the house with Sethe, Paul D finds that his way of coping is not so effective after all. While "nothing in the world" can break into the defenses Paul D has built to keep his own painful past at bay, something out of this world—that is, Beloved, the spiteful ghost demanding revenge—succeeds in opening the little tin box, leaving him broken apart, unable to overcome the tyranny of the painful, agonizing memories that are the legacy of slavery.

Sethe and Paul D seem to have very little choice either about Beloved's presence among them or about the intrusive past, relentlessly squeezing out the present and blocking the possibility of the future for Sethe as she beats it back, for Paul D as he seals it off. Further complicating Sethe's struggle with the past is her determination to protect her daughter: "To Sethe, the future was a matter of keeping the past at bay. The 'better life' she believed she and Denver were living was simply not that other one. . . . As for Denver, the job Sethe had of keeping her from the past that was still waiting for her was all that mattered" (42). Sethe's effort to keep the past at bay proves futile, for Denver compulsively repeats its stories. Unlike a murder mystery that unravels what has happened, this novel focuses on the preconditions of a murder and gains suspense by raising questions about whether the characters will extricate themselves from their pasts to find "a way out of this no way" (95).

There is nothing in the novel to suggest that Paul D and Sethe can find that way without the intervention of external forces—perhaps the force of a supportive community. It was, after all, the collaboration of the black community with the conditions of slavery that led to the murder in the first place. The day before the slave catchers arrive, Baby Suggs presides over an abundant feast; the next day, the very neighbors who had enjoyed the event wake up with changed hearts, envious of Baby Suggs's power and relative affluence: "Too much, they thought. Where does she get it all, Baby Suggs, holy? Why is she and hers always the center of things? How come she always knows exactly what to do and when? Giving advice; passing messages; healing the sick, hiding fugitives, loving, cooking, cooking, loving, preaching, singing, dancing and loving everybody like it was her job and hers alone" (137). It is this envy—and the Fugitive Slave Act of 1850, which made interference in the capture of a runaway slave a felony—that causes members of the community to keep quiet when strange whites—always potential slave catchers—appear in town. Some deliberately choose not to send out a warning in time for Sethe to flee; others apparently directly betray her: "Six or seven Negroes were walking up the road toward the house: two boys from the slave catcher's left and some women from his right" (148).

Years later when Stamp Paid interferes in Sethe's life by showing Paul D the newspaper account of the murder, he is still blaming Baby Suggs for giving up her public life after the murder. Eventually, however, he too feels defeat and begins to understand that his old friend gave up because her own community had betrayed her and because "whitefolks had tired her out" (180): "to acquire a daughter and grandchildren and see that daughter slay the children (or try to); to belong to a community of other free Negroes—to love and be loved by them, to counsel and be counseled, protect and be protected, feed and be fed—and then to have that community step back and hold itself at a distance—well, it could wear out even a Baby Suggs, holy" (177). The consequences of Stamp's and the community's betrayal of Baby Suggs are still operating in 1874 when Sethe is trying to take Baby Suggs's advice and "*lay it all down, sword and shield*" (173). By telling Paul D about the murder, Stamp perpetuates his original betrayal, paving the way for Paul to desert Sethe just as Stamp had cast off Baby Suggs.

After a lifetime of struggling first against the evils of slavery and then the racist atrocities that followed emancipation, Stamp begins to feel himself succumb, as Baby Suggs did, to terminal fatigue. Although slavery was abolished some ten years before the opening scene of the novel, many of its atrocities are still being practiced. Stamp recalls, "Eighteen seventy-four and whitefolks were still on the loose. Whole towns wiped clean of Negroes; eighty-seven lynchings in one year alone in Kentucky; four colored schools burned to the ground; grown men whipped like children; children whipped like adults; black women raped by the crew; property taken, necks broken. He smelled skin, skin and hot blood. The skin was one thing, but human blood cooked in a lynch fire was a whole other thing" (180). While emancipation ended slavery and rendered the Fugitive Slave Act meaningless, it opened the door to the widespread slaughter of former slaves.

One day on the river Stamp reaches under his boat for what he believes is a red cardinal's feather, only to pull up "a red ribbon knotted around a curl of wet woolly hair, clinging still to its bit of scalp" (180). The final straw is not so much the "people of the broken necks, of fire-cooked blood," but rather of "black girls who had lost their ribbons" (181). In spite of the despair he feels when he finds the red ribbon connected to a fragment of human scalp, Stamp does not, in the end, give up; rather he acknowledges his own role in spurning his old friend and contributing to Sethe's isolation. And after eighteen years, he sets in motion the events that eventually lead to the restoration of community. Stamp's first step is to visit Ella, one of the community of women who has turned against Sethe. Stamp confesses his own guilt to Ella, setting her up to muster the women of the community to exorcise the ghost.

The cause of the tragedy of Sethe's murder of her own child, then, is slavery itself and the public policies—the Fugitive Slave Act and lynching—that slavery engendered. Contributing to the tragedy, however, is the black community's failure to stand in unity against a common enemy.

In the context of the final climactic scene when the women of the town—some presumably part of the "throng . . . of black faces" who rejected Sethe (152)—come back to claim her, the title of this provocative novel gains significance. The epigraph of the novel is from a passage in Romans. Paul is explaining why God has chosen the Gentiles over the Jews. But curiously, Paul is quoting Hosea. Of Hosea's three children, representative of the Israelites temporarily rejected because of their own betrayal, one is called "not beloved." After a period of retribution, God reclaims the lost people, or community of Israelites, symbolically renaming the children. Hence the words from Romans:

> I will call them my people,
> which were not my people;
> and her beloved,
> which was not beloved.

In both the Old and the New Testament versions, God claims for himself a rejected people, restoring community in Hosea and establishing community in Romans.[14]

After Sethe purchases the word *Beloved* with ten minutes of stand-up sex with the engraver of headstones, she regrets not having spent another ten minutes for *Dearly*, so that on her baby's headstone would have been "every word she heard the preacher say at the funeral" (5). There were, of course, other words spoken at the funeral, and the absence of those words from Sethe's consciousness reflects their absence from the social fabric: "Dearly Beloved, we are gathered together . . ." are the words commonly associated with the marriage ceremony, a rite Sethe was denied in slavery. Rather than gathering in support and love, the community had already joined in betraying Sethe. What is missing from her life is the gathering together of black people for mutual protection and mutual goals.

The word *beloved* has special significance in the struggle for racial justice. Perhaps the first use of the phrase in African-American revolutionary writing was in David Walker's 1829 *Appeal.* The son of a slave, who lived as a free black in Boston, Walker boldly and persuasively advocated the overthrow of slavery by violent insurrection. He addresses his readers in the preamble to the *Appeal* as "My dearly beloved Brethren and Fellow Citizens" and elsewhere in the document as "beloved Brethren." Martin Luther King, Jr., John Lewis, and other civil rights leaders punctuated their speeches with references to the "beloved community." Alan Paton's powerful novel of 1948, *Cry, the Beloved Country*, like *Beloved*, memorializes the past while calling for a different future from the one that history has predicated. *Beloved*'s affirmation of the power of gathering together in the spirit of community as the necessary step in breaking the tyranny of the violent past is at the heart of this lyrical, poetic novel that is fixed in the prosaic reality of history. Like Nel's "fine cry" at the end of *Sula*, this novel's "circles of sorrow" are contained in a profound cry for a country and community that is "not beloved."

The "beloved community" consists not just of those who gather at metaphorical rivers today, but those who have contributed their lives and their stories to history. In addition to revealing the role of the community in breaking the back of vengeance, the story reconnects contemporary readers who may be cut off from their historical past with the powerful community that history preserves. A few months after the publication of *Beloved*, Lerone Bennett, speaking at the Martin Luther King, Jr., Center for Nonviolent Social Change, warned that King's dream of a "beloved community" is threatened, that violation of community leads to chaos, and that community is not a gift, but "a never-ending struggle." *Beloved* invites its readers to rejoin and to revive the struggle for community and, in its climactic scene, condemns violence as an adversary of community.

While *Beloved* is an appeal for the recovery of the black community, it raises questions about the possibility of responsible cooperation with whites. Denver, literally born on the road to freedom, is obsessed with what to her is a happy story, the tale of her own birth and the strange white woman who delivered her. When Sethe is fleeing slavery and is no longer able to walk, she lies down in the forest and prepares to die. The daughter of an indentured servant, Amy Denver, is running from poverty toward a better life in Boston when she responds to Sethe's groans and for no particular reason begins to care for her (32–34). For two days Amy fights for the suffering slave woman's life and that of her baby. After Amy delivers the baby in the bottom of a boat on the Ohio river, the two women work together:

> On a riverbank in the cool of a summer evening two women struggled under a shower of silvery blue. They never expected to see each other again in this world and at the moment couldn't care less. But there on a summer night surrounded by bluefern they did something together appropriately and well. A pateroller passing would have sniggered to see two throw-away people, two lawless outlaws—a slave and a barefoot whitewoman with unpinned hair—wrapping a ten-minute-old baby in the rags they wore. But no pateroller came and no preacher. The water sucked and swallowed itself beneath them. There was nothing to disturb them at their work. So they did it appropriately and well. (84–85)

Years later, Sethe remembers how Amy tended her wounds and massaged her mutilated feet, all the time explaining that "Anything dead coming back to life hurts" (35) and that nothing can "heal without pain" (78).

There are references to Amy Denver and to Denver's birth throughout the novel. In the first scene when Sethe explains to Paul D that she would never have made it without the white girl's help, he says: "Then she helped herself too, God bless her" (8). Named for the white girl who saved her life, Denver is "a charmed child," destined for survival (40). When Paul D expresses concern about Denver, Sethe explains: "Nothing bad can happen to her. . . . Even when I was carrying her, when it got clear that I wasn't going to make it . . . she pulled a whitegirl out of the hill" (42).

Perhaps because of Amy, Denver expects whites to be helpful. When Denver finally leaves the house where she has been confined with her mother for so long, she goes first to a black woman and then to two white people who have played an important role in Sethe's life. A brother and sister who once worked for abolition and aided runaways, Miss Bodwin and her brother Edward are still kind to former slaves, though they no longer have the sense of mission that once fired their lives. Like Rufel in

Dessa Rose, and many abolitionists after the war, they helped the slaves to freedom but took no responsibility for what happened thereafter.

Denver does find that Miss Bodwin, at least, is willing to help her. Apparently bored and in need of a new cause in the post-war doldrums, she takes on Denver as an experiment, teaching her "book stuff" (266), and preparing her to attend Oberlin College, the first coeducational college in the country, which as early as 1835 was admitting black students. But Denver also encounters undeniable racism and exploitation. The Bodwins require that Janey, their black servant, work night and day even though she has a family of her own to care for. When Denver knocks on the front door, it is Janey who suggests that she must learn "what door to knock on" (253). Leaving by the back door, she sees a container in the form of a kneeling black boy, his mouth wide open, stuffed with coins to pay for deliveries; on its pedestal are the words "At Yo Service" (255).

Edward Bodwin, unlike his sister who still creates projects for the future, looks back at his days as an abolitionist with sadness: "Those heady days were gone now; what remained was the sludge of ill will; dashed hopes and difficulties beyond repair. A tranquil Republic? Well, not in his lifetime" (260). Now more than seventy, Bodwin has given up his public identity and fallen into a kind of reverie for his personal past, longing for childhood objects—toy soldiers and a watch chain—that he buried long before in the yard of the house now occupied by Sethe and her daughters. Yet, when he looks back to his days as a reformer, almost twenty years after he first aided Sethe, he feels the loss: "Nothing since was as stimulating as the old days of letters, petitions, meetings, debates, recruitment, quarrels, rescue and downright sedition. Yet it had worked, more or less, . . . Good years, they were, full of spit and conviction" (260).

Edward Bodwin's feelings might well have been expressed by a disillusioned reformer in 1987 looking back to the heady days of the civil rights movement. For white liberals, after the civil rights movement, as after the Civil War, there is no excitement, no drama, and no heroism in the day-to-day struggle to insure the survival of what has been won. Emancipation, like the Civil Rights Acts of 1964 and 1965, created only the possibility, not the actuality, of racial justice. But then, as after the movement, it was not clear how to create that just society that is still more dream than reality.

Unknowingly, Bodwin has one more role to play in Sethe's tragic struggle with Beloved. On his way to give Denver a ride to work, he approaches the house where Sethe has lived for eighteen years as a tenant.

Sethe, suddenly gripped by the past, rushes forward. The scene is emblematic: a crazed black woman raising an ice pick to kill a white man; a crowd of black women simultaneously intervening to prevent violence and falling back away from the rising white man; and the ghost Beloved vanishing from sight, from the characters' lives, and from the narrative. The very moment that Sethe lets go of Beloved, she and Denver rush toward the crowd of women who have gathered to release this grip of the past: "Away from her to the pile of people out there. They make a hill. A hill of black people, falling. And above them all, rising from his place with a whip in his hand, the man without skin, looking. He is looking at her" (263).

To interpret this passage in narrow terms would be to reduce its power in the novel. The temporal context of the episode—the summer of 1874—invites readers to consider it as a foreshadowing of the Hayes Compromise of 1877, which assured the political dominance of white males and the falling back of masses of blacks. The image is ambiguous enough to suggest parallels with our own time when the liberation of individuals—from economic, social, and political conditions imposed by the past—is still played against the backdrop of a multitude "of black people, falling" and a white man "rising from his place with a whip in his hand." The "sixty million" who died in slavery to whom the novel is dedicated have their counterparts in the 1980s, among the homeless, the unemployed, the incarcerated, the drug addicted, the mad.

Sethe's release from the past seems to have been caused by a series of events, each qualifying what follows. Sethe first turns her destructive impulses away from herself to Edward Bodkin—the 1870s equivalent of the white liberal—whom she has mistaken for a slave catcher. Intervening to prevent violence are two women from different generations: Denver, now grown up and ready to start her own life, and Ella, who managed to escape her own catastrophic past. Helping them is the community of women who instead of fostering violence now prevent it. The white man, threatened by violence, responds in kind. Yet, the struggle that is community, the novel seems to suggest, cannot be sustained by violence.

In a society that often buries its own history, narratives that process, embrace, and contain the past may play a part in imaginatively reconnecting past, present, and future. An episode that recalls contemporary racist horrors is the one in which Stamp Paid fishes from the river a red ribbon still attached to a piece of a child's scalp. For some readers this episode may evoke images of the multitude of black bodies that have been found floating in rivers since the Civil War, the most notable being that of young

Emmett Till, whose story Morrison, the year before publication of *Beloved*, dramatized in a play, *Emmett Dreaming* (1986). Others may be reminded of a series of murders of black children in Atlanta, some found in rivers. The final pages of the novel include an account of Paul D's experiences in the Civil War, which connects his story with the final episode of the civil rights movement: "After a few months on the battlefields of Alabama, he was impressed to a foundry in Selma along with three hundred captured, lent or taken coloredmen. That's where the War's end found him, and leaving Alabama when he had been declared free should have been a snap" (268).

By connecting this story through place names with the movement and with contemporary ordeals endured by the black community, the novel evokes the stream of history flowing from the terrors of the slave ships, atrocities committed in the name of the Fugitive Slave Act, and the violence that followed in the wake of the nonviolent civil rights movement, reminding readers that some who walked away metaphorically from the Selma campaign, far from finding their way made easy, may have encountered instead, like Paul D, a series of violent acts against blacks: "Freeing yourself was one thing; claiming ownership of that freed self was another" (95).

Typical of Morrison's mode in her earlier novels, she leaves unanswered questions: Does the life-draining Beloved disappear because the cycle of violence has at last been broken? In breaking that cycle, does the community become beloved? Does the task of freeing oneself from the catastrophic past require a communal act?

There is also considerable ambiguity about whether Paul D will succeed in healing Sethe and about what kind of future Sethe and Paul D will have. By leaving readers in the middle of a restorative process, Morrison involves us in these questions as well. After Beloved disappears, Sethe, lying in bed with no inteniton of getting up, tells Paul D that she has "no plans" (272). Slowly he prepares to bathe her and rub her feet, repeating the restorative process that Amy Denver began so long before. Determined to "put his story next to hers," he woos her away from what appears to be terminal despair by appealing to the future: "me and you, we got more yesterday than anybody. We need some kind of tomorrow." Countering Sethe's long held belief that her children are the best part of her, he gently touches her face, saying, "You your best thing, Sethe. You are" (273). Sethe's response, "Me? Me?" leaves readers to consider how these two—and their counterparts today—might make a future together.

In an interview in the *Washington Post*, Morrison observes that, compared with the accounts of the life of the historical Margaret Garner, the original for Sethe, *Beloved* is actually a happy story. Sethe is merely haunted by the past and the ghost of the child she killed to prevent her from being returned to slavery. Margaret Garner, who killed her child for the same reason, was tried and then sent back to her master.[15] But for all its rendering of the agonizing past, *Beloved* is a happy story for other reasons. Sethe and Paul D have finally come to terms with the past and have moved into the possibility of a future together. Denver, long confined to the house with her mother, finally takes responsibility for herself and seeks help from others. She finds a community of people who offer her work, education, and even love. Like Milkman and Guitar at the end of Morrison's *Song of Solomon*, the characters are poised in possibility.

Beloved seems to defy those two romantic notions about the past that underlie much social and political conservatism—one expressed in Jay Gatsby's ill-fated dream that we can "relive the past," and the other in Nick Carraway's fatalistic conclusion that we cannot step out of it since we are like "boats against the current, borne ceaselessly into the past." It is, of course, such assumptions that allow the recurrence of racial violence and the quiet resegregation of our schools. While dramatizing the power of the past—and vengeance—to dominate the present, *Beloved* also suggests ways to transcend the tyranny of that past. The way Morrison tells her story invites readers to connect their lives with the catastrophes of historical narrative and to engage in the creation of a community that is truly beloved. By forcing readers to confront the most painful aspects of the slave past and by narratively breaking the cycle of violence that recreates ever-new versions of that past, *Beloved* is a challenge for our time.

But Sethe's story is not a parable with a paraphrasable content, nor does it offer practical political proposals. Rather, it evokes and suggests parallels between her time and our own. By 1855, when an estimated sixty thousand slaves had made it to the limited freedom of the North, efforts to reduce their number were bolstered by the Fugitive Slave Act. After emancipation, the freedom of former slaves, like that some African Americans have enjoyed since the passage of the civil rights acts, was threatened by the retrenchment of the federal government away from progress toward racial justice, as well as by fragmentation within the African-American community.

As the first volume of an intended trilogy, the narrative reaches a kind of stasis rather than a conclusion. Each of the three main characters has broken free from the past. Denver for the first time in her life is being approached by a young man, and she turns to him with "her face looking like someone had turned up the gas jet" (267). Paul D is ministering to Sethe, feeling what it means to have "a woman who is a friend of your mind," and is determined to "put his story next to hers" (273); and Sethe herself is responding to his tenderness. Since Morrison has said that the next volume in the trilogy will be set in the Harlem Renaissance, we can anticipate that future novels will move forward in time to other generations fighting their own battles with public and private history, and that if they are consistent with her five published novels, they too will challenge readers to become part of the narrative process.

Beloved does not have a final word, in that its last word *Beloved* is also the first. By coming full circle, Morrison embraces and contains not only her terrifying story, but the troubling questions it raises. Although in the climactic scene, the characters seem to have transcended the cycle of violence that has defined their lives, the narrative voice declines to comment on this outcome, leaving readers to enter the story again and again in search of clues that might resolve its ambiguities and answer its questions. That never-completed process is where this novel lives.

CHAPTER TWO

FROM THE GREAT WAR TO WORLD WAR II

When Toni Morrison published her first novel, *The Bluest Eye*, in 1970, four years after *Jubilee*, the civil rights movement was already history. Much of the progress the movement would bring had already come: desegregation in schools and public institutions was underway; legislative goals had been achieved. But the spirit and leadership of the nonviolent movement were gone. Martin Luther King, Jr., was dead, and SCLC, the organization that supported his work, was floundering. In 1966, militants had replaced moderates in two movement organizations: Stokely Carmichael had taken over the leadership of SNCC from John Lewis, and Floyd McKissick had replaced James Farmer as the head of CORE. Both organizations were fraught with conflict by the end of the sixties. While the surviving leaders of the nonviolent movement were struggling to establish a viable agenda, militants called for "black power," and some endorsed violence as a valid means of achieving their goals.

What a few years before had been hailed as the victories of the movement had, in fact, done little to change the lot of African Americans living in ghettos. The past six summers had seen major urban riots, and the escalation of the Vietnam War had further contributed to the fragmentation of the movement as activists alternately took stands against the violence in Southeast Asia, for the violence at home, and for and against diverting attention from the domestic struggle to the antiwar movement. Richard Nixon had begun to undermine the achievements of the movement and to turn around those mechanisms—court-ordered school busing and the fair-housing enforcement program—intended to achieve desegregation. Lyndon Johnson's promises of a Great Society had long rung hollow, and those who attended to the condition of African Americans were struck with the powerlessness of the young and impoverished who had hardly been affected by the civil rights movement. While long-term reformers were asking "what next?" others, particularly the young and disaffected, were declaring "so what!"

Into this climate of despair and failing energies came *The Bluest Eye*, focusing on the helplessness of a deprived, abused, and finally discarded child. The primary time of the action is 1940–41, twenty years after the peak of the Garvey movement; some ten years after factionalism had begun to divide the NAACP over issues of elitism, accommodation versus separatism, and leadership; shortly after Ella Baker began what would be a long-term commitment to community organization for racial and social justice; and the very year that A. Philip Randolph's call for a mass demonstration in Washington persuaded Roosevelt to issue an executive order prohibiting discrimination in the defense industry while racism still prevailed in the military.

Like Pecola Breedlove, the protagonist of *The Bluest Eye*, Toni Morrison grew up in Lorain, Ohio, and was ten years old when the events of the novel take place. Like Claudia MacTeer who narrates much of Pecola's story, Morrison had resourceful parents who worked hard and facilitated her journey into the mainstream of society. Claudia's mature narrative voice may not be very different from the one Morrison herself would use to relate events from her own somewhat similar childhood, anchored in the same history. Claudia's voice speaking from the perspective of a future time, perhaps the late 1960s when the novel was written, passes judgment on her younger self, and by implication on others like her, for contributing to the decline of those less fortunate.

By 1970 the numbers of African Americans who had graduated from prestigious universities, enrolled in professional schools, and taken positions of responsibility had grown considerably since the early days of the movement; at the same time, life for those left behind who were either unemployed or earning meager wages in oppressive and insecure jobs had hardly improved. Many who had so recently profited educationally and economically from the movement, like Claudia MacTeer, must have been plagued by a sense of responsibility for those left behind. *The Bluest Eye* spoke powerfully in 1970 of the gap that had developed between those whose lives the movement had improved and those for whom change had come too late.

The Color Purple, published in 1982, relates events in the personal lives of its characters from the beginning of the twentieth century through World War II, but its climactic events are set in the first half of the 1940s, about the time Alice Walker was born. Unlike in *The Bluest Eye*, published some twelve years earlier, the time of the setting has little to do with the thematic thrust of the novel, and as Bernard Bell has observed, *Purple* is more appealing as a folk romance than as a realistic document of representative blacks fulfilling a commitment to improve their lot.[1] To appreciate the limitations of an ahistorical narrative in which characters change as a consequence of psychological or moral forces without reference to socioeconomic or political realities, we have only to imagine what would happen to the happy-ever-after feeling at the end of *The Color Purple* if, for example, by moving the characters into the world of historical consequence some had been destroyed by the Depression or war. While Celie can dismiss the events reported in newspapers as "crazy," some characters in the more historically rooted novels know that what happens in their own lives is significantly conditioned if not determined by economic and social conditions that are part of the society, as well as by events that are reported in the news.

The handling of history—or its neglect—relates not only to the internal structures and themes of the two novels but to the contemporary audience, the supposed reader at the time of publication. By using a narrator who looks back nostalgically to 1941, Morrison links past and present, suggesting parallels between the temporal setting and the time of publication; by freezing her narrative in the voices of undated letters, Walker removes it from historical specificity into the realm of "once upon a time." The acceptance of Celie's sexual relationship with Shug and her success as an entrepreneur seem much more compatible with 1980s California, where Walker wrote the novel, than with rural southern Georgia, where she grew up. But unlike Alice Walker, the characters in *The Color Purple* never do anything to initiate social change, nor is there a narrative voice with moral authority speaking, like Claudia in *The Bluest Eye*, of the responsibility that privileged people have for the oppressed. Nor is there an omniscient overvoice to connect the story to the post-movement period in which the novel was first conceived and read. Nettie, who once thought of her life as a missionary as part of a larger movement to help raise black people everywhere, seems to have relinquished that dream by the time she leaves Africa at the end of the novel.

The struggle to create a structure and an agenda for progress in civil rights was well underway in 1941, the narrative time of *The Bluest Eye* and one of the climactic years of *The Color Purple*, but the gains achieved by A. Philip Randolph, Ella Baker, and others were small compared to the enormity of the task ahead. The 1930s had seen considerable controversy among those in the black community who were committed to the cause of civil rights; leaders within the established organizations were at odds with one another about goals and priorities, and few could agree on what forms progress would take. For African Americans who had little evidence that life could be different, the concept of progress had little meaning.

Yet the impact of *The Bluest Eye* and *The Color Purple* at their times of publication depends on some evidence that their characters progress, as in fact they do, in ways that are consistent with the spirit of the time in which the books are being read. While the degeneration of Pecola Breedlove informs the narrative thrust and focus of *The Bluest Eye*, the possibility of the development of social responsibility and moral fortitude lives in Claudia's voice as she looks back on her past, regretful of her contribution to that tragedy. Claudia has progressed from a thoughtless child to a mature adult with a stinging social conscience. Presumably something has happened in the society and in history to bring about such a change. Progress in *The Color Purple*, however, is entirely in the private domain, as characters relinquish destructive behaviors, develop satisfactory personal relationships, and construct economically viable lives in the midst of a hostile, oppressive, and essentially unchanged society.

The Bluest Eye

The Bluest Eye is timeless as a study of a social outcast (Pecola) who rejects herself and of a one-time outsider (Claudia) who climbs into the mainstream of society and in the process renounces those she left behind. It is also a timely novel in its treatment of the tragic consequences for those who attempt to escape blackness by an unrealistic and self-destructive yearning for the symbols (blue eyes) and privileges of whiteness or by renouncing their historical identity in order to accommodate to the white world. By condemning the choices of both Pecola, who languishes in madness, and Claudia, who eventually prospers, the novel leaves readers in the wake of civil rights legislation with the task of discovering what it means to succeed in a white, still racist, but somewhat more open society.

The attempt to escape slavery and its painful consequences memorialized in *Beloved* in the late 1980s was, in its way, the subject of Morrison's first novel, *The Bluest Eye*, published seventeen years earlier, as indeed it has always been the "subject" of black lives in America. As in *Beloved*, the primary narrative is set in Ohio, and its background episodes are set in the South. The main action takes place in 1940–41, however, with flashbacks to the previous two decades. In *The Bluest Eye*, as in *Beloved*, Morrison explores the impact that social conditions have on the private acts of individuals and provokes her readers to confront the historically determined social parameters within which individuals create their private lives. Both of these novels explore the conditions that lead a parent to commit an irreparably destructive act against a child. While Sethe's murder of her baby girl is a "rough response to the Fugitive Slave Act," Cholly Breedlove's rape of his daughter, Pecola, is his response to the "guilt and impotence" imposed by the racist society that declares him worthless (127).[2]

As in *Beloved*, the progressive narrative of *The Bluest Eye* is periodically interrupted by lengthy accounts of past events, and though it focuses primarily on Pecola, it is not only her story. Like *Dessa Rose*, *The Bluest Eye* is told in different voices: in the sophisticated voice of a narrator who observes the characters and comments on their actions with the detachment of a sociologist or psychologist; in Pauline Breedlove's thoughts as she recalls her childhood, her love for Cholly, and the loss of that love; in Pecola's mad ruminations and fantasies about the rape that serve to block out the horror of her life; and in the repentant adult voice of Claudia MacTeer, looking back at the time when she met Pecola Breedlove and watched a community destroy her. Morrison first composed this novel in the third person and later added Claudia to serve as "a bridge" between Pecola and the reader.[3]

The four parts of the novel—"Autumn," "Winter," "Spring," "Summer"—are each constructed from fragments of these various voices. The opening fourteen lines of the novel are taken intact from grammar-school readers featuring Dick and Jane and the paraphernalia of middle-class white America: "Here is the house . . . See Father. He is big and strong . . . Bowwow goes the dog . . . Here comes a friend . . . Play, Jane, Play" (7). Serving as an ironic contrast to the limited world of the black characters, these images from the idealized world of white suburbia are evoked again in excerpts used as epigraphs preceding the narration of the various traumatic incidents in Pecola's life. The epigraphs form an imagistic compressed version of the story: Put out of their "house," Pecola's family moves from being outsiders to being "outdoors"; a child humiliates Pecola and accuses her of killing his "cat"; her "nice" mother shames her in front of her friends and lets her know she prefers the little Dick-and-Jane white girl whom she takes care of; a sadistic child molester tricks her into poisoning a dog; her "big and strong" father rapes her; her friend Claudia betrays her.

The novel focuses on two households: the squalid storefront apartment where Cholly and Pauline Breedlove live with their two children Sammy and Pecola and the more spacious, though still shabby, house where Mr. and Mrs. MacTeer live with their daughters Claudia and Frieda, their boarder, Mr. Henry, and for a brief time Pecola, who comes there as a foster child. The narrative moves in and out of other households in the black community, as it relates Pecola's visit to three aging prostitutes; her intrusion into the pristine rooms of a middle-class, light-skinned black woman; and her visit to the back-room apartment of the eccentric Soaphead Church, who, like Celie in *The Color Purple*, writes a letter to God.

Subjected to that most dreaded condition of impoverished communities, being "outdoors" with "no place to go" (18), Pecola lives in a private world that is virtually untouched by change. Claudia, on the other hand, is an outsider who becomes an insider by joining the public world that creates what we call history. Pecola and other outsiders—in 1941, in 1970, when the novel was published, in the 1990s and beyond—are casualties of the historical and public arena; by her own admission, Claudia and her counterparts then and now are silent collaborators in keeping the Pecolas of the world "outdoors."

Claudia's voice frames the narrative and interrupts periodically to relate her own experiences and to interpret the story she tells. She looks back with sorrow and regret from a mature perspective to that time in her own life when she was obsessed with the suffering of little Pecola Breedlove, whom she both pitied and shunned. Narrating the opening episode of "Autumn," Claudia relates how Pecola came to live in the MacTeer household as a foster child. Comparing herself to that pathetic little girl, Claudia, who rejects and mutilates her white dolls, cannot understand Pecola's infatuation with the blue-eyed, baby-doll image of Shirley Temple or her disturbing desire to know how "you get somebody to love you" (29). Obsessed with her own ugliness and with an overwhelming desire to be white, blue-eyed, and beautiful, Pecola concludes that such beauty would bring her love. As she moves through her version of Pecola's story, Claudia acknowledges the role that the African-American community played in sealing Pecola's doom by adopting white standards of beauty. She takes responsibility for having acquiesced in Pecola's destruction, insuring that "the damage done was total" (158). Although there are no episodes in which she is overtly cruel to Pecola, Claudia feels guilty. Her sins are those of omission, of acquiescing to the oppressor's values and reinforcing Pecola's sense of self-hatred.

The Bluest Eye, like *Beloved*, is interpretatively historic, both in its public and private aspects. Firmly placed in time and place, it explores both the social conditions and personal choices that led up to the "what happened." In the opening pages Claudia reveals that Cholly raped his daughter, that she became pregnant, that Cholly and the baby are dead, and that the narrative that follows will explain "how," not "why," it happened. "Our innocence and faith were no more productive than his [Cholly's] lust or despair. What is clear now is that of all of that hope, fear, lust, love, and grief, nothing remains but Pecola and the unyielding earth. Cholly Breedlove is dead; our innocence too. The seeds shriveled and died; her baby too. There is really nothing more to say—except why. But since why is difficult to handle, one must take refuge in how" (9).

The "how" involves going back to Cholly's infancy and childhood. Long passages of third-person narration relate the events in his past—his mother throwing him on a junk heap when he was four days old, his aunt rescuing him and raising him until she died when he was thirteen, white men interrupting his first sexual experience to mock and terrify him, his father rejecting him for a crap game, his meeting and marrying Pauline. Passages of third-person narration alternating with italicized accounts of Pauline's own thoughts reveal a romantic girl with a deformed foot who dreamed of a stranger who would come and take her away from the small Kentucky town and "lead her away to the sea, to the city, to the woods . . . forever" (90).

When Cholly meets Pauline Williams in the 1920s, he is trapped in a system established by slavery and in many ways unchanged by the Civil War, a system that doomed most rural blacks to unemployment or to low-paying jobs as domestics and sharecroppers. When Cholly escapes the South and settles in Morrison's own home town of Lorain, he expects a better life: "In that young and growing Ohio town . . . which sat on the edge of a calm blue lake, which boasted an affinity with Oberlin, the underground railroad station, just thirteen miles away, this melting pot on the lip of America facing the cold but receptive Canada—What could go wrong?" (93).

But even in Ohio in the 1920s and 1930s, things went wrong. Cholly and Pauline encounter indifference, poverty, unemployment, and the Great Depression. They have two children, Pecola and Sammy. The romance fades from their marriage, Cholly begins to drink and abuse Pauline, and she neglects her own children to concentrate on becoming the "ideal servant" for a respectable white family (100). Cholly's abuse of his family eventually lands him in jail and Pecola in the MacTeer household as a foster child.

Compared to the Breedloves, the MacTeer family is relatively privileged. Their house is large enough to take in a foster child and a boarder. There is enough food and plenty of fresh milk for the children. Though they are part of "a minority in both caste and class," moving about "on the hem of life," they either hang on or eventually "creep singly up into the major folds of the garment" (18). Early in 1941, after Pecola has been returned to her own family, Cholly in a drunken state rapes and impregnates her, precipitating her descent into permanent madness. In a desperately futile attempt to save Pecola's baby, Claudia and her sister Frieda plant marigold seeds as an offering to God, believing that, when the flowers bloom, Pecola will somehow be saved. Looking back on her own childhood and adolescence, Claudia considers the conditions that led to the waste of a human life, acknowledges her personal responsibility for not preventing Pecola's decline, and recognizes the part that public factors play in that process.

Meditating on the way that her own memory becomes confused with her mother's memories, Claudia relates the story her mother told about being swept up with her hand still on her hip by a tornado that destroyed much of Lorain but set her down safely: "The anticipation and promise in her lolling hand are not altered by the holocaust. In the summer tornado of 1929, my mother's hand is unextinguished. She is strong, smiling, and relaxed while the world falls down about her. So much for memory. Public fact becomes private reality, and the seasons of a Midwestern town become the *Moirai* of our small lives" (146). The story of Mrs. MacTeer's composure in the midst of a tornado serves as a parable for interpreting Claudia's relation to the storm that destroys Pecola and others like her, as she recognizes in later life her role in that destruction. Public facts—poverty, racism, and unemployment—condition the private lives not only of that storm's direct victims, but also of those who imagined that they have escaped and whose composure may have contributed to and certainly did nothing to alleviate the suffering of others.

By acknowledging her responsibility with shame and regret, Claudia implies that she has overcome the tendency to use others as she once used Pecola; at the same time, her voice is a challenge to readers of 1970 both to be alert to the public facts in their time, which are inevitably molding their private realities, and to recognize those aspects of private experience that in turn inform public life. In the last half of the 1960s when this book was composed, as in 1970 when it came to light, American society was caught up in a different kind of storm. To maintain a state of "composure" during those years was to be indifferent to the public facts affecting the lives of millions of Americans in Vietnam, in the riot torn inner cities, and in the rapidly disintegrating civil rights organizations. By 1970, the agenda of the movement had shifted away from King's nonviolent protests, and those advocating black power, black pride, and the black aesthetic had moved into the forefront of the struggle for racial justice. Quieter, less dramatic efforts, such as voter registration movements, were not receiving media attention. Many African Americans were publicly rejecting the middle-class values and accommodation to white society that Claudia regrets having embraced. From the vantage point of mature adulthood, Claudia confesses her guilt: "We were not free, merely licensed; we were not compassionate, we were polite; not good, but well behaved. . . . We substituted good grammar for intellect . . . rearranged lies and called it truth" (159).

As she interprets her role in Pecola's decline, Claudia becomes the moral register of the novel. Calling attention to her guilt in feigning compassion, morality, and integrity, the novel challenges readers in 1970 and beyond, readers who probably use correct grammar and live well-behaved middle-class lives, to examine their own attitudes and behaviors to determine whether they have concluded, as Claudia did, that the Pecolas of the world have "no right to live" (160).

The Pecolas of the world do not even have exclusive rights to their own life stories. All of the information about Claudia comes from her narration, whereas Pecola's story and that of her parents emerges from the third-person narration, from Claudia's reminiscences, and from Pecola's own stream of consciousness. Except for a few episodes from her childhood and the points where her life intersected with that of Pecola, we know little about the details of Claudia's life. *The Bluest Eye*, then, is not about how Claudia made it, but it is in part about how she is responsible for those who did not. Like Harker in *Dessa Rose*, however, Claudia may not have done more than "feel bad" about those who did not make it.[4]

In 1970, when this novel was published, the black-is-beautiful movement was at its height. By exploring the forces that create a young black girl's image of herself as ugly and unlovable, *The Bluest Eye* contributed to the growing awareness at the time of the damage inflicted on black children by a culture that exalts the white aesthetic. The story of Pecola's obsession with whiteness and her intense desire for blue eyes is set exactly during the time that psychologist Kenneth Clark's research into the damaging effects of the white aesthetic on black children was a public issue. The discovery that many African-American children in 1940 envied white children and, like Pecola, thought of them as more beautiful and more desirable than themselves made a significant contribution to the progress of civil rights.

In the spring of 1940, Clark and Phipps published the results of their now famous doll test.[5] By showing black and white dolls to children and by asking about their perceptions of the dolls and themselves, the researchers concluded that black children in segregated schools saw themselves as inferior and accepted that inferiority as reality. When, in 1950, Clark repeated the test in South Carolina's Clarendon County to determine whether the original findings applied to children in the South, the original results were confirmed. He found that most black children presented with the two dolls and asked which they like the best will choose the white doll but recognize the black doll as most like them. The most disturbing finding to him was the degree to which some children accepted their perceived inferiority as reality. One boy, selecting the black doll as like himself, said, "That's a nigger. I'm a nigger." The results of these tests later became key evidence in the case against segregation that led to the defeat of the separate but equal doctrine in *Brown v. Board of Education* in May 1954.[6]

Pecola's conclusion that people with white skin and blue eyes are superior is reinforced by her mother, who early in her marriage spends much of her time at the movies, where she becomes infatuated with white women's beauty and so disdainful of her own race that all she can see is her daughter's "ugly" blackness. When Pauline deflects attention and love from Pecola to the "little pink-and-yellow girl" she cares for (87), she permanently fixes in her daughter's mind the notion that love is reserved for little white girls with blue eyes. Pecola's sense of worthlessness and her desire to have blue eyes is further reinforced by her squalid living conditions in the inadequate space of a store front: "They lived there because they were poor and black, and they stayed there because they believed they were ugly" (34).

Pecola repeatedly encounters people who confirm her belief that blackness is a curse. Noticing that the white shopkeeper who sells her candy is reluctant even to touch her by taking money from her hand, she concludes that the "distaste must be for her, her blackness" (42). Further contributing to Pecola's decline is the prejudice that divides not only blacks and whites but members of the black community from each other. The light-skinned Maureen Peel briefly befriends Pecola, then turns on her, yelling, "I am cute and you ugly! Black and ugly" (61). When a schoolmate lures her into his "pretty gold-and-green" house, his "pretty milk brown" mother, feeling her middle-class status threatened, throws Pecola out, calling her a "nasty little black bitch" (75–76). Middle-class brown-skinned women look down on their darker neighbors; black boys fawn over light-skinned girls. No one takes any realistic steps to help the dark-skinned and doomed Pecola; except for Claudia and Frieda, everyone seems to take pleasure in the scandalous news that Cholly Breedlove has impregnated his own daughter.

Reacting differently to a world that embraces the white aesthetic, Claudia despises with "unsullied hatred" the white, blue-eyed baby doll her parents give her every Christmas (19); she recalls what it was like to hate not only the dolls but to transfer "the same impulses to little white girls" (22). Later in life, she concludes in her adult voice, the shame she feels for such hatred leads her to camouflage it with a show of love, which soon grows real: "The best hiding place was love. Thus the conversion from pristine sadism to fabricated hatred, to fraudulent love. It was a small step to Shirley Temple. I learned much later to worship her, just as I learned to delight in cleanliness, knowing, even as I learned, that the change was adjustment without improvement" (22).

Recalling an episode in which a "high-yellow dream child" (52) harasses Pecola, Claudia concludes that her own hatred of the light-skinned and therefore more privileged blacks is misdirected anger. "The *Thing* to fear was the *Thing* that made *her* beautiful, and not us" (62). In the mature voice in which she narrates the story of her relationship with Pecola, Claudia regrets not just her failure to help her friend but the self-deception that allowed her to embrace the values that contributed to Pecola's destruction.

After the rape and a series of other traumatic events, Pecola becomes convinced that she, in fact, does have blue eyes: "A little black girl yearns for the blue eyes of a little white girl, and the horror at the heart of her yearning is exceeded only by the evil of fulfillment" (158). Cooperating in her final leap into madness is Soaphead Church, a child molester, who comes from a long line of people who have conducted their own quest for blue eyes, separating themselves "in body, mind, and spirit from all that suggested Africa" (132). For generations, his family had "married up," choosing mates for the whiteness of their skins, thus "lightening the family complexion and thinning out the family features" (133). The consequence of such selective breeding has been the "weakening of faculties and a disposition toward eccentricity" (133). After Soaphead tricks Pecola into killing an old dog he hates by promising her blue eyes, Pecola loses her last hold on reality.

The main events of the novel—the rape, Pecola's decline into madness, and the birth and death of the baby—take place in the months immediately preceding the entry of the United States into World War II—a time of growing awareness of the damage done to black children by a society that equates beauty and whiteness and a time when the system that reinforced that view was just beginning to change.[7] When Claudia and Pecola were approaching puberty as the war was escalating, climbing into the mainstream of American life had begun to be possible for blacks, but it was still an awesome task. Claudia MacTeer made it; Pecola Breedlove never had a chance. Claudia acknowledges that she and others like her who have managed to rise above their origins use the Pecolas of the world to bolster their own sense of belonging in the mainstream: "We were so beautiful when we stood astride her ugliness. Her simplicity decorated us, her guilt sanctified us, her pain made us glow with health, her awkwardness made us think we had a sense of humor. Her inarticulateness made us believe we were eloquent. Her poverty kept us generous. . . . We honed our egos on her, padded our characters with her frailty, and yawned in the fantasy of our strength" (159). The culprits in the crimes against Pecola, then, are not just the social conditions that destroyed first her parents and then Pecola herself, but those within the black community who use the less fortunate to facilitate their own success in a racist society.

Claudia's final words attack her earlier acquiescence to human suffering:

> I talk about how . . . it was the fault of the earth, the land, of our town. I even think now that the land of the entire country was hostile to marigolds that year. The soil is bad for certain kinds of flowers. Certain seeds it will not nurture, certain

> fruit it will not bear, and when the land kills of its own volition, we acquiesce and say the victim had no right to live. We are wrong, of course, but it doesn't matter. It's too late. At least on the edge of my town, among the garbage and the sunflowers of my town, it's much, much, much too late. (160)

Looking back on this attitude of her childhood, Claudia still seems torn between two positions. On the one hand, she sees that conditions were such that some flowers would not bloom—that no Pecolas had a chance for life at that time in America. On the other hand, she acknowledges her role in assuring that the social soil would remain barren. The attitudes that condemned Pecola both elevate and obstruct Claudia, who finds success by embracing the values that condemn the less fortunate. By metaphorically taking on the values and aesthetics of the blue-eyed world and turning her back on her own history, Claudia, and presumably others like her, have entered the mainstream of American life and, the novel suggests, paid dearly for their privileges. In this context, the social and political conditions that condemned such families as the Breedloves also exacted a price from those who rose into the middle class like the MacTeers.

By beginning and ending the novel with Claudia's mournful words, full of regret for her part in the suffering of others, Morrison invites her readers—presumably mostly middle class and relatively privileged—to consider their own role in perpetuating such suffering. The civil rights movement may have opened doors for middle-class blacks and made it possible for others like Claudia to climb from the fringes to the center of American society, but what, Morrison's readers might wonder in late 1970, after almost two years of Richard Nixon, had it meant for people like Cholly, Pauline, and Pecola Breedlove? Just as Morrison memorializes those whom slavery doomed in *Beloved,* here she dramatizes the pain of those who are still enduring its consequences today.

In a published "conversation" with Gloria Naylor, Morrison speaks of the time in the late sixties when she was writing *The Bluest Eye:* "I was primarily interested in the civil rights movement. . . . It was my time of life also. . . . If the best thing happened in the world and it all came out perfectly in terms of what the gains and goals of the *Movement* were, nevertheless nobody was going to get away with that; nobody was going to tell me that it had been that easy. That all I needed was a slogan: 'Black is Beautiful.'"[8]

In her first novel, Morrison seems to be addressing the black middle classes, urging an understanding of and respect for the black underclass, white readers who wonder why it may be necessary to insist that black *is* beautiful, and all who avoid considering their own role in maintaining a social system that ignores and destroys the weak and unfortunate.

The Color Purple

Implicit in *The Bluest Eye* is the assumption that, in 1970, blacks who long for the blue of blue eyes, randomly scattered by God and accessible only to whites, are betraying their racial identity and the possibility of achieving more worthy goals. In Alice Walker's *The Color Purple*, however, the surprising patches of pleasure-giving purple scattered by God from time to time in an otherwise dull landscape are gifts, not goals, that tell the characters that, even if they are poor and black and female, there is pleasure in the world, there for the taking—sometimes.

There are striking similarities in the content of *The Bluest Eye* and *The Color Purple*. The climactic events of both novels are set in the early 1940s; both have male characters who abuse women and a protagonist who is a victim of incestuous rape resulting in pregnancy. An account of the rape is featured in the opening passages of each book. Both focus on black characters with rural Southern origins who are outside the mainstream of society. In *The Bluest Eye*, a child molester writes a letter to God, explaining the conditions that led him to fondle little girls; *The Color Purple* begins with the letter Celie has written to God relating how she was sexually abused by a man she believes to be her father. But Morrison and Walker use this subject matter to create very different narratives.

The Bluest Eye, except for two relatively conventional flashbacks late in the novel, is a more or less sequential narrative with dates that suggest the real events which impinge on its private, fictional world. By noting in the first pages that the main events take place in the fall of 1941, the novel announces its strong historical underpinning. Characters refer to Roosevelt and the Civilian Conservation Corps (CCC) camps (24). They mention Greta Garbo and Ginger Rogers (17), Shirley Temple (19) and Claudette Colbert (57), Dillinger (46), Hoover (45), and Henry Ford (24). Though Morrison's novel may resemble *Dessa Rose* in its use of multiple narrators, the primary narrative—the story of how Claudia becomes involved with Pecola Breedlove and feels responsible for what happens to her—is straightforward and linear. Like Margaret Walker's more conventional historical novel, *Jubilee*, *The Bluest Eye* is historical in its content, chronology, causality, and intended objectivity, and perhaps in its thematic impact on readers of 1970. In its third-person passages and in Claudia's relatively impersonal, evaluative first-person narration, it includes in its field of vision the larger society that either nurtures or condemns private lives.

The stories that emerge in *The Color Purple* are mediated entirely through the intensely subjective form of personal letters.[9] Almost half are written in Celie's voice over a thirty-year period and are addressed not to a person living within history, but to that wholly extra-historical external entity, God. Because Celie's husband hides from Celie the letters written by her better educated, more worldly sister, Nettie, a missionary in Africa, Nettie's letters do not enter the novel until years after they are written. Though in her letters Nettie refers indirectly to history by relating her experiences outside the South and in other countries, the fact that Celie reads them all at once and only after many years have passed takes them out of time; and in the end, Nettie abandons her mission in the greater world, in this way coming around to Celie's position that what is reported in newspapers is "crazy" (178)—that history is irrelevant and personal life (and pleasure in things purple) are what it's all about. Such a pleasure principle, the theme of the 1980s "me-generation," paved the way for *The Color Purple* when it came on the scene, preaching in its fashion, and perhaps inadvertently, private or "free enterprise" to Reagan's America. The private nature of its epistolary form highlights its theme of personal satisfaction. Like *Dessa Rose*, its subjective voices propose private solutions to what may be public problems, just as the objective form of *The Bluest Eye*, fashioned from the analytical, distanced voices of social critics, reflects its public, societal concerns.[10]

Alice Walker's *The Color Purple* is the story of two sisters whose lives take widely divergent paths but who, in spite of a thirty-year hiatus in communication, manage to keep their devotion to each other alive. Celie remains in an isolated part of rural Georgia. She marries a man who does not love her, raises his children, cooks and keeps house, plows his fields, endures his always-unwelcome sexual assaults, and even cares for and eventually learns to love his long-time mistress, Shug Avery. Nettie, on the other hand, gets an education, finds work helping a minister and his wife (Samuel and Corrine), and goes with them to Africa as a teaching missionary. Most of the fifty-odd characters, however, are part of Celie's small world. Among them are her mother, who dies young and leaves her at the mercy of Pa, who has already raped her, fathered two children by her, and taken those children from her; her husband, Albert, whom she calls "Mr. ______" until the end of the novel, who beats and abuses her until his mistress Shug Avery intervenes; her stepson, Harpo, and Harpo's women, Sofia and Squeak; and Shug herself, who becomes first Celie's friend and then her lover.

Nettie's private world in Africa is even smaller than Celie's. Except for Samuel and Corrine, and Celie's children, Adam and Olivia, whom they have adopted, Nettie is close to only two other people: Catherine, a tribal woman, and Catherine's daughter, Tashi, who eventually marries Adam. Nettie's private life runs parallel to but on a separate track from historical time. Though she travels widely, reads newspapers, has tea with bishops and even a head of state, and struggles to prevent exploitive land developers from destroying the traditional tribal life of the Olinka, Nettie ultimately has no perceivable impact on history, nor does it have any significant influence on her. The novel's final indifference to historical contingencies results in obvious as well as subtle anachronisms. For example, when Nettie arrives in Liberia in what seems to be the early twenties, she meets with President Tubman—who did not, in fact, take power there until 1944.

Celie's early letters to God focus on her feelings and the daily lives of the few people she knows. Written when she was an adolescent and during the years she is married to Albert, the letters have a timeless, ahistorical quality, as she writes about the details of her daily life without ever referring to the world outside the rural central Georgia farming community that is home. Although she records the passing of time—"it took him the whole spring, from March to June" (11)—she never notes the day of the week or the year. Episodes that must have occurred during the Depression do not include any references to poverty or hunger; in fact, there always seems to be more than adequate, even abundant food. Celie never mentions public officials, historical events, or even the local welfare office. She does not report hearing news on the radio or talking to people from other parts of the country, and though she does notice Pa "rattling his paper" (12), she gives no indication that she knows what newspapers might contain.

Years later, when Shug lies in bed reading the paper out loud to her, Celie dismisses the news: "People fussing and fighting and pointing fingers at other people, and never even looking for no peace" (178). When Celie writes about Shug's experiences on the road, she never refers to the places Shug has been or to anything Shug might have learned from the greater world, though she does boast that Shug knows "everybody . . . Sophie Tucker . . . Duke Ellington" (95). Until Celie finally receives a letter from Nettie announcing that she is "coming home before the end of another year" (100), the characters in Celie's world give no indication that they are aware that the world extends beyond the nearby Georgia towns. Toward the end of the novel, Celie does begin to show an interest in some of the folk beliefs of the Olinka, but she reports them as curiosities (231–33). In fact, Celie lives the first five decades of her life in a private world in an isolated part of rural Georgia, and she cannot even imagine that the world outside could affect her. When one character comments on the difficulty of proving something "to the world," Celie stops short, realizing that she has never thought about the world: "What the world got to do with anything, I think" (52).

Unlike any other novel considered in this study, *The Color Purple* does not contain a single date. *The Bluest Eye*, for example, which covers approximately the same time period, alerts readers to the exact date of the climactic events in the beginning pages—"Quiet as it is kept, there were no marigolds in the fall of 1941"—while the passages narrated in Claudia's adult voice sustain the historical perspective as she looks back at and comments on the past. Since events in the Georgia scenes of *The Color Purple* are related entirely by Celie, who shows little awareness of time passing, there are few clues to the historical context of the private events that she relates.

There is, however, a deducible chronological order to events. We do know that Celie is fourteen when she first gets pregnant, that she is "near twenty" (10) when she marries, and that Harpo, her stepson, who is twelve at the time Celie marries, is eighteen or so when he marries Sofia, with whom he has five children. She leaves him, has another child, and then soon after is arrested for "sassing the mayor's wife" (75). Sofia spends some eleven-and-a-half years in jail or in service to the mayor's wife before she is released to go back to her family; she had known Harpo some twelve years before her sentence began (77). References to the United States becoming involved in World War II suggest that Celie left Albert and moved to Memphis with Shug around 1941, when she was approximately fifty. The final episodes of the novel occur before the war is over, as Celie receives word that Nettie was on a boat "sunk by German mines" (216). Shortly afterward Celie receives a letter, presumably written just before Nettie left Africa, noting that "nearly thirty years have passed" since she has heard a word from Celie (217). Celie was twenty then.

None of the public events of the time, a time which we can date from the events of Celie's private life, enter her consciousness or the novel. Celie must have been born around 1890, when most of the progress that former slaves made after the war had been reversed. The first decade of Celie's life, a time when the cooperation of Northern liberals and former abolitionists with Southerners in reinforcing institutionalized racism culminated in the Supreme Court decision of *Plessy v. Ferguson* (1896), which insured segregation with the doctrine of "separate but equal." Celie is first raped by her stepfather around 1903–4, a time when Charles W. Chestnutt noted that racial prejudice was "more intense and uncompromising" than it had been since the end of the Civil War;[11] and W. E. B. Du Bois was declaring in *The Souls of Black Folk* that "the problem of the Twentieth Century is the problem of the color line." In the years that follow, when the NAACP and other organizations were laying the foundation for reform, neither Celie nor any of her friends seem to be aware of that activity. Though she reports that Shug and Albert read newspapers, she never mentions the names of the papers—whether, for example, they read *The Crisis* or other publications that recount the activities that eventually lead to reform.

Celie lives some five decades in that period before World War II when many black Americans never expected that there would be an end to the racism and exploitation they endured. She lives through the Depression in a rural society where for most African Americans daily survival was a struggle, yet she never reports suffering its consequences. Not once does Celie register the dramatic and accelerating events that would eventually change the way African Americans live.

Though Celie may not be aware of the historical context of her own life, she repeatedly relates events that grow from racist practices which limit the lives of the men and women who are her family and friends. The first episode involving white people occurs when she is fourteen and pregnant. Her stepfather, who has taken her out of school, is cleaning his gun when "a bunch of white mens" carrying guns come walking across the yard. Then, Celie reports: "Pa git up and follow 'em. The rest of the week I vomit and dress wild game" (12). Celie is clearly not aware of the racist and sexist practices that are affecting her life—that she was taken out of school to help out during hunting season, one of the few ways rural black women had to earn extra money.

Years later, Celie writes more directly of encounters with racism, such as the time the mayor's wife stops Sofia in the street, admires her children, and then asks if she would like to be her maid. Sofia refuses and starts a fight that ends in a brutal beating and a twelve-year jail sentence. Her friends play Brer Rabbit by sending Squeak to convince the warden that Sofia is much happier working in the prison jail as a laundress than she would be working as a maid. While the trick works, the prize is not exactly the briar patch: The warden rapes Squeak and condemns Sofia to live in the mayor's home and to raise his children. Years later, after she has completed her twelve-year stint as a twentieth-century slave, she returns to her own family. The mayor's daughter, Eleanor Jane, brings her baby on a visit to Sofia and playfully begs her to praise the child. But Sofia balks, explaining that she does "not love Reynolds Stanley Earl" (224), who will no doubt learn to exploit her and other black people. Though Celie reports these events, she does so without giving any indication that she is aware of public policies and practices. In most southern states, including Georgia, convicts were leased to individuals to serve as unpaid laborers. While Celie would have no way of knowing whether money was exchanged in this case, she does not seem to understand that under the convict lease system many black people suffered such a fate.

Most of the characters do not seem particularly bothered by the racist insults they endure. Even though she sits in the front seat of the car when she teaches the mayor's wife how to drive, Sofia gets in the back without an argument when Miz Millie explains that she will never see "a white person and a colored sitting side by side in a car" (90). When the otherwise assertive Shug Avery arrives for a visit, she reports without apparent rancor that she and her new husband Grady have "been driving all night. . . . Nowhere to stop, you know" (93). Writing to Nettie about the trip to Memphis, Celie simply reports that "us have to find a road going off into the bushes to relieve ourselves" (175); she observes that, when Shug is on the road, there is "no place hardly to stop and really wash herself" (179).

In other novels considered in this study—Morrison's *Sula* and Childress's *A Short Walk*—characters experience profound humiliation or engage in outrageous acts of rebellion in the face of such discrimination, but although Celie and Shug complain about the inconvenience, they do not seem particularly disturbed that they cannot stop at gas stations with restrooms or in hotels to shower and sleep.

Nettie, on the other hand, is bothered by the Jim Crow train she took to New York:

> We had to ride in the sit-down section of the train, but Celie, there are beds on trains! And a restaurant! And toilets! . . . Only white people can ride in the beds and use the restaurant. And they have different toilets from colored.
>
> One white man on the platform in South Carolina asked us where we were going. . . . When we said Africa he looked offended and tickled too. Niggers going to Africa, he said to his wife. Now I *have* seen everything. (113–14)

Early on, Nettie writes of her work as a contribution to "the uplift of black people everywhere" (115). In subsequent letters, she not only expresses concern about racism—and sexism—but reveals that she has communicated these concerns to the children. When she explains to Olivia that the Olinka people "don't believe in educating girls," the child concludes that they are "like white people at home who don't want colored people to learn" (133).

Although she has taught the children about racism in America, Nettie worries that they will be shocked at "the hatred of black people" and will not be able to "manage the hostility towards them" (218). Like Claudia in *The Bluest Eye*, Nettie acknowledges the harm that comes from color prejudice within the black community, and she worries about Adam's wife, Tashi, who has very dark skin. Having read American magazines and noticed that blacks do not "truly admire blackskinned people like herself . . . especially . . . blackskinned black women," Tashi fears that, since she is very black and has "scarification marks on her cheeks," Adam will be attracted by light-skinned women and desert her (235). To convince her of his love, Adam scars his own cheeks.

As a child, Nettie played the role of teacher to her older sister. But after Celie leaves school, Nettie has little success, since Celie, so preoccupied with her private world, cannot make sense of her sister's lessons. When, as a pregnant fourteen-year-old girl, Celie listens to Nettie trying to explain about the discovery of America and how the earth is round, she only pretends to understand: "I just say, Yeah, like I know it. I never tell her how flat it look to me" (12). Nettie's letters, written over a period of some thirty years, are the history book that may help Celie discover "what the world got to do with anything" (52), how her own life relates to past events, how the society she lives in has shaped that life, how another society has shaped her children, and how private family history is an aspect of cultural and racial history.

Nettie takes care to point out, on the one hand, that there were accomplished black people living in great cities thousands of years ago and, on the other, that some Africans captured and sold other blacks into slavery "because they loved money more than their own sisters and brothers" (111). She describes Harlem during its heyday, where black people "live in such beauty and dignity" (114); discusses the British role in the slave trade (117); and tells of the founding of Liberia (117), of the customs of the Olinka (137), and of the "*mbeles*" or forest people, who refuse all contact with whites (193). Other letters include stories about the founding of Spelman Seminary, W. E. B. Du Bois, and King Leopold of the Congo.

Nothing in Celie's letters indicates that she learns from or even registers Nettie's lessons in public history; she does, however, respond to the private family history. Although Celie learns about some of the outrageous forms of racism through Sofia's experiences, her knowledge is fragmentary, without cause or cure, without historical vision or substance. It is not until she reads what Nettie has written about her own family's past history that Celie realizes the persistence, range, and historical context of racism. Her real father, she learns, was a "well-to-do farmer," who opened a store and prospered. The white merchants resented a black man taking away some of their business and had him dragged from his home and lynched. Celie's mother, who had one child (Celie) and was pregnant with her second (Nettie), went mad when neighbors brought his mutilated and burnt body home. When she remarried, her new husband not only abused her but raped Celie and fathered two children by her. Pretending that those children belonged to his ailing wife, the man gave them to Samuel and Corrine, who adopted them and hired Nettie as their nurse.

Until she receives Nettie's letter relating this history, Celie has assumed that her mother's husband was her own as well as her children's father. When she finally learns the truth of her personal history, Celie begins to take control of her life. Like Soaphead Church in *The Bluest Eye*, she chides her divine correspondent for inattention to the welfare of children: "You must be sleep," she writes to God. In a very narrow and personal way, Celie gains a sense of history, of her own past, and of how events she knows nothing about have had a great deal to do with her after all.

The dawning of a limited historical consciousness is the impetus that leads Celie to act. When she goes to her stepfather to learn more about her past, he explains that Celie's father died because he "didn't know how to get along" with whitefolks and that to get along and to get ahead in a racist society "you have to give 'em something. Either your money, your land, your woman or your ass." He chose "to give 'em money" (155). Like Randall Ware in *Jubilee* and Grange in *The Third Life of Grange Copeland*, he has concluded that no white man can be trusted; like Ware, he chose to buy off the white man. And it is Celie's and Nettie's inheritance that he uses to insure his security and prosperity.

The two women in Celie's life teach her quite different lessons: Nettie writes of a life of hardship and sacrifice serving others, and Shug professes a life of self-indulgence. While Nettie has given much of her life to raising Celie's children and ministering to a backward tribe in Africa, Shug has abandoned her own children and ignored the hardships of other blacks. Although she generally treats Celie with kindness, Shug readily, though temporarily, abandons her when she meets an appealing young man. Shug teaches Celie to support herself, to enjoy luxury, and to live for the moment in a narrowly personal, if not totally selfish, world. Nettie, on the other hand, who regrets "the hurt we cause others unknowingly" (158), teaches her to be concerned for the welfare of others and to live in communal and historical awareness. Shug espouses the pagan, Nettie the Christian way of life.

The novel's epistolary form does not allow for clarity about how Celie has been shaped by these two conflicting points of view, but the scales are weighted on Shug's side, since Nettie comes home to a private world, having abandoned her life's work with a feeling of failure. Shug's voice has authority that others lack, and her words, always mediated through Celie's, are doubly validated. She has a way of speaking that causes others to accept what she says and does, and she often gets the last word. Although Albert has repeatedly abused and beaten Celie in the past, Shug is able to make him stop. When Shug teaches her about sex, Celie listens and learns. When, after a long road trip, she announces to Celie and Albert that she is married to Grady, they both welcome her back on her own terms. After Grady has left and the two women are living together, Shug persuades Celie to wait for her while she indulges in what she imagines will be her last fling with a young man. Celie, even though she is plagued by jealousy, concludes that Shug has "a right to look over the world in whatever company she choose" and that love should not infringe on such rights (228).

Independent, competent, and powerful, Shug controls the lives of others, convinces Celie that she can earn a living making trousers, and directs her friend Mary Agnes's career as a singer. It is Shug who articulates the theological concepts which give the novel its title. Writing to Nettie about the notion that what God likes most is admiration, Celie quotes Shug: "I think it pisses God off if you walk by the color purple in a field somewhere and don't notice it. . . . People think pleasing God is all God care about. But any fool living in the world can see it always trying to please us back" (167). Consistent with Shug's view is Celie's final apostrophe to God—"Dear stars, dear trees, dear sky, dear peoples. Dear Everything. Dear God" (242).

Not only does Celie accept the gospel according to Shug, but Nettie, living in Africa and having no contact with Shug, seems to discover it on her own. And since Corrine died and she married Samuel, Nettie has found the satisfaction in her personal life that she never experienced in her public commitment. For Shug, God is neither an "old white man" nor an impersonal force (166); rather, God exists "inside everybody" and is primarily concerned with personal feelings, pleasing people, and gaining "admiration" (167). Having abandoned the idea of a divine being with a human form, Nettie arrives at an internal deity that "each person's spirit is encouraged to seek . . . directly" (218). When she returns from her fling, Shug demands and gets affirmation as Celie reassures her that she and Albert, whom she can at last call by his name, have become friends, and that all they do is talk about how much they both love her. In terms of the novel, what Shug *says* is what *is*.

There are difficulties in trying to evaluate this novel and to understand the significance of both the historical and ahistorical reality it portrays and the kind of statement it made to the reading public in 1982, which received it with accolades—and the Pulitzer Prize in fiction. Though *The Color Purple* points out the evils of racism and sexism, its strongest voice —mediated and validated through Celie—is that of Shug, whose self-indulgence and pleasure seeking lead her to use and discard people at will. Through the authority of Shug's behavior and voice, the novel seems to endorse an essential selfishness, to say, "enjoy yourself however you can, even if you hurt people who love you." Though at the end of the book Celie has work and money and is reunited with her family, she is still tied emotionally to someone who values her own independence more than Celie's happiness and who dominates her, just as Albert had done when they were married, though in a more acceptable way. Considered in this light, Celie's progress seems severely limited, not so much by social evils, as by the nature of her private emotions and attachments.

By celebrating lovemaking and money-making in 1982, *The Color Purple* seems to appeal to those on the make in Reagan's America and to gloss over the economic and social realities of many blacks; by dramatizing the need for blacks to feel pride in blackness, *The Bluest Eye* had a particular historical validity in 1970 and beyond. *The Color Purple* imagines a kind of miraculous transformation of personality, nurtured by powerful though incomplete personal love, rather than the slow, tedious change that is the result of hard work and commitment to making the most of life. Of Walker's four novels, *Meridian* and parts of *The Temple of My Familiar* portray characters changing slowly in the context of a public commitment, while the transformations of character in *The Third Life of Grange Copeland* and *The Color Purple* take place in the personal domain.

To understand why some readers have had difficulty reconciling the scenes set in Georgia with the African scenes, it is helpful to stop and consider how Celie responds to her sister's letters. Although she rejoices at the news that her sister and children will soon be returning to Georgia, she expresses no concern about their future or about how such educated, experienced people will adjust to life in a racist society in an isolated part of Georgia. She never asks her sister questions or expresses concern for her suffering, nor, as we have seen, does she give any indication that Nettie's history lessons have affected her. She mainly records her personal transformation as she breaks free from the shackles of marriage to an abusive man.

Readers have no clues as to what relationship there might be between Nettie's letters and Celie's growing confidence and competence, whether Celie's capacity to take charge of her own life has been strengthened by a sense of her own past conveyed in Nettie's letters, by Shug's love for her, or by a combination of both. The absence of any reference in Celie's letters to a connection between her internal world and the external history-bound society is key to this novel's place in its own time. Everything we know about Celie we know from what she chooses to write—first to God and then to Nettie. Except when she receives the news of her own family history, her father's murder, and her mother's decline into madness and finally death, Celie does not respond to Nettie's stories of the past. Nettie, who never receives Celie's letters, must continue to write in a vacuum. By having Celie grow only in personal ways while remaining oblivious to Nettie's history lessons, the novel never confronts the complex interaction between public and private experience.

At the end of the novel, Celie has inherited land, a house, and a store. She is an independent clothing designer and seamstress, and she employs others to help her manufacture pants and to work in the store. She is a successful capitalist. By the time Nettie returns, Celie has settled into a comfortable existence in an entirely private world. Although she reports that Sofia's sons have gone to war and that Eleanor Jane's husband has stayed home, she does not know what the war is about or seem to care to find out where "France, Germany or the Pacific" might be (222).

The Color Purple, then, is the story of Celie's journey from total ignorance and isolation to awareness that she lives in a complex world with people in perpetual conflict; of her movement from oppression to personal freedom, even license; of Nettie's journey from youthful enthusiasm to despair over her inability to deal with conflicting social and economic forces; and of both women's discovery of a satisfying—though perhaps in Celie's case, nonexclusive—personal love. But it is also a story with a happy ending that some readers may have difficulty believing. With the exception of refusing to tolerate certain patronizing behavior from whites, Celie does not seem interested in doing anything about the injustices she has encountered, but rather she settles for her work and the personal relationships that make up her world. Somehow the conflicts and animosities that have marred the lives of Celie and her friends have been resolved, as each of the characters has found some way to tolerate, accept, even love the others. The only character who is excluded from the final scene of reconcilation is Grady, an unattractive man who abuses women and deals drugs. The other once-abusive males are tamed, domesticated, and integrated into the community. The final scene, then, has a happy-ever-after sense of closure.

Readers who dare to speculate about what is likely to happen to the characters after the emotion-charged reunion at the end of the novel may find their imaginations inadequate, however, since there is no ambiguity about the ending. The world may be at war, whites may continue to oppress blacks, and individuals may suffer and lead crippled lives; but Celie and Nettie are united again, Celie is a prosperous capitalist, and all is right with *their* world. When, at the end of the novel, young Henrietta asks why they always have a family reunion on July 4, Harpo explains that since "white people are busy celebrating they independence from England . . . black folks don't have to work . . . [and] can spend the day celebrating each other" (243). Mary Agnes responds by teasing Harpo: "I didn't know you knowed history" (243). Public history, then, is something that happens to white people; black people must create their own separate and mainly private history.[12]

The ending of the novel seems to suggest that, in the early 1940s, only a private happiness was possible for Celie and others like her and that a similar retreat to private pleasures may have been appropriate in the early 1980s. While *The Bluest Eye* calls attention to the emotional cost of rising above one's origins and succeeding at the expense of others, *The Color Purple* seems to validate a quest for private happiness, as characters succeed in private endeavors (Celie), enjoy newly available pleasures (Shug), and relinquish their youthful idealism (Nettie). There is nothing in the novel to suggest that the characters might take action to initiate any political or communal action.

Midway through the novel, after she learns that Albert has been keeping Nettie's letters from her, Celie wants to kill him, but Shug, who argues that "nobody feel better for killing nothing" (122), convinces her to begin sewing to pass the time and to deal with the anger: "Times like this, lulls, us ought to do something different" (124). Shug is referring to the lull in their sexual activity caused by Celie's anger, but readers in the early 1980s might apply the statement to the public arena as well. The early 1980s, after all, were a kind of lull in political activism, a time when activists might have felt that they would have to wait—do something different—before changes could occur. Some readers might have concluded that the 1980s were a time for enterprising individuals to profit, as Celie does when she begins to make pants.

The Color Purple does not suggest what will follow the lull for Celie or her counterparts today. It rather affirms a time of reconciliation and consolidation within the black community without proposing a program, plan, or forward thrust. For readers in the early eighties, it affirmed the regrouping of the private world that is personal, intimate, and even sexual, one that is free from political and social concerns. At that time it was a novel for Yuppie and Buppie America, an America that was a bit tired, a bit bewildered, driven to cut losses, consolidate gains, and cultivate its own private, profitable gardens.

While Celie's new-found happiness is at least in part dependent on inherited property and success as a capitalist, others in Celie's shoes would not likely have a wealthy lover like Shug to stake her to a business; nor would many, or any, such abused and disadvantaged women expect to inherit a comfortable house and a profitable store. Though she defends the outcome of Celie's story, Walker has acknowledged both that the inheritance of property and a profitable business "is not a viable solution" for the poor and that there is a need to search for "better solutions for the landless, jobless, and propertyless masses."[13]

Speculation about the conditions that led Walker to create for Celie a happy ending that is unambiguously materialistic and personal or why she chose to end her novel a decade before the events that would make it likely for Celie to find outlets for any kind of social action leads to intriguing possibilities. Perhaps the stories novelists tell are so conditioned by the receptiveness of the culture that even a writer like Walker—who for more than twenty-five years has been committed to social action on many fronts, including the civil rights movement, the women's and the antinuclear movements—inadvertently speaks to the values of the audience dominant at the time she composes a novel, in this case an audience listening for reassurance that seeking economic prosperity and personal gratification are valid enterprises.

Six years intervened between the publication of Alice Walker's first and second novels and another six years between her second and third novels; seven years after readers of *The Color Purple* first marveled at its magic, she has spun a very different tale. *The Temple of My Familiar* may have been generated by a shifting political consciousness, one that is growing less comfortable with a public ethic of self-indulgent, personal satisfaction. Or, it may be that the vast popularity of *The Color Purple* freed Alice Walker to write a novel that is more consistent with her larger political agenda than is the private vision of *The Color Purple.*

CHAPTER THREE

HARBINGERS OF CHANGE: HARLEM

From its early days as the mecca of African Americans, Harlem incorporated many domains, including the underworld of organized crime; the demimonde of petty hustling, gambling, prostitution, and speakeasies; the society of jazz and blues musicians; the professional community of middle-class doctors, lawyers, teachers, and ministers; the painters, sculptors, and photographers who contributed powerful visual images to the larger communal projects of racial uplift and social reform; writers who created a rich and varied literature, ranging from the refined, elitist poetry of Countee Cullen to the down-home tales of Zora Neale Hurston; the group of interpreters like Alain Locke who promoted the art of the time and with it the idea of a Renaissance; and finally the activists and intellectuals—Marcus Garvey, James Weldon Johnson, W. E. B. Du Bois, Jesse Fauset, and A. Philip Randolph—whose efforts to reform the lives of African Americans were rooted in Harlem and at the same time pointed beyond it, whether to Africa or to a new society based on fundamental social change.

The events that led to the establishment of a dominant African-American population in Harlem began at the turn of the twentieth century, and by the mid-teens that process had taken hold. Madame Walker, the wealthy cosmetologist, moved to Harlem in 1914, a time when entrepreneurs were aggressively promoting Harlem real estate and the dream of building a distinguished Negro community. That year the first of the major theaters in the area desegregated; in the months and years to follow, the major black churches, newspapers, and civil rights organizations opened offices in Harlem. By the mid-teens Scott Joplin and other accomplished musicians had joined the migration to Harlem. By the mid-twenties, painters like Aaron Douglas were integrating their own more abstract artistic concerns with the specific social agenda of Harlem's activist community.

In 1917, Marcus Garvey's Universal Negro Improvement Association was founded there. Social reformers, writers, and intellectuals were already engaged in the controversies and inquiries that would feed the creative activities, as well as the political turmoil, of the next decade. Characteristic of this unique period was diversity within unity. The geographical boundaries of Harlem provided the common context for a disparate group of African Americans with competing world views to interact. At odds were Garvey's emergent black nationalism, A. Philip Randolph's socialism, and W. E. B. Du Bois's eclectic radicalism; but even among these most famous leaders of often conflicting schools of thought there was overlap and common ground. Du Bois, for example, was alternately labeled a socialist, an integrationist, an elitist, and a radical. Hardly anyone stayed in a fixed intellectual stance; ferment, controversy, and change carried the day.[1]

The writing community, stimulated and nurtured by controversy, was rapidly producing a rich and varied literature. Within a few years came Claude McKay's *Harlem Shadows* (1922) and *Home to Harlem* (1928), Jean Toomer's *Cane* (1923), Jesse Fauset's *There Is Confusion* (1924), Langston Hughes's *The Weary Blues* (1925), Nella Larson's *Quicksand* (1928), and Wallace Thurman's *The Blacker the Berry* (1929). During this time of optimism and confidence, many writers believed that poetry and fiction were appropriate agents of social change. For Zora Neale Hurston, contributing to that process meant studying the black culture and preserving its folklore; for Jessie Fauset, it meant portraying the trials of genteel, middle-class blacks. While Langston Hughes in his poetry and fiction embraced the broad spectrum of African-American experience, some writers like Countee Cullen openly defended the black writer's involvement in art for its own sake. But this rich period of fertile interaction among such diverse members of the Harlem writing community was short-lived. By 1935 when Zora Neale Hurston published *Mules and Men*, the literary movement that both fed and was nurtured by the Harlem Renaissance was essentially over.[2]

The Harlem Renaissance stood approximately equidistant from the end of the Civil War and our present time. Some sixty-five years had intervened between the war and the end of this enormously fertile period when such writers as Du Bois, Frazier, and others saw themselves as "instruments of history-making and race-building."[3] But the self-assurance and heady aspirations that fed the productivity of the period were shattered by the Depression. Harlem was a place where dreams and a vision of an ideal society were suddenly crushed by external, public events.

What was true for successful Harlemites was true for other visionaries of the day. But Harlem was unquestionably a unique community where in a limited space and within a relatively short time—approximately a decade and a half—artists, writers, intellectuals, and entrepreneurs converged with events to create an environment where they could examine the past, assess the present, and envision a future; and out of those endeavors came lasting sociological, literary, and artistic manifestations of those visions. What is perhaps most striking about the Harlem Renaissance is that its end was definitive. Of course the Depression caused devastating effects across the social spectrum and undoubtedly many individuals and institutions never recovered; but Wall Street did recover, and for most privileged people, so did the American way of life. But the decline that the Depression initiated in Harlem has continued to this day, and for some six decades, residents and visitors to what one of Louise Meriwether's characters calls those "mean streets" have been continually struck with reminders of how private lives are affected by public policy.

When Meriwether published *Daddy Was a Number Runner* in 1970, the civil rights movement had come to almost as precipitous an end as had the Harlem Renaissance in the early 1930s. Measured from Garvey's ascendance on one end to the Depression on the other, Harlem's golden age lasted approximately as long as the heyday of the movement extending from *Brown v. Board of Education* in 1954 to the death of Martin Luther King, Jr., in 1968. The diverse factions within the movement were informed by positions similar to the competing world views of the Harlem Renaissance. In both periods that diversity was contained until outside forces—economic disintegration, urban riots—intervened.

Just as the Depression had undercut and dispersed the dream of a great black community that was Harlem, so had riots, assassinations, and the Vietnam War scattered the energies of those participants in the civil rights movement inspired by the dream of a "beloved community." In both periods, the end seemed sudden, and yet the modern civil rights movement was nurtured by ideas and individuals that came of age during Harlem in its heyday. When A. Philip Randolph came to Harlem as a young man in 1911, he began making public speeches on the street at the intersection of

135th Street and Lenox Avenue;[4] he gave his last major public address at the Lincoln Memorial some fifty years later, immediately before introducing Martin Luther King, Jr., whose celebrated "I Have a Dream" speech has informed all memories of that day. Considered by many to have been the highlight of the movement, the massive March on Washington on August 28, 1963 was orchestrated by A. Philip Randolph and Bayard Rustin, both old Harlemites. Planning the march from a small office on West 130th Street, Randolph and Rustin conceived of the event as complementary to King's street demonstrations in the South.[5]

Daddy Was a Number Runner is set in the depths of the Depression, when the hunger and degradation of many Harlemites contrasted dramatically with the vision of Adam Clayton Powell, Sr., of his community as "the Promised Land to Negroes everywhere."[6] By setting her novel in Harlem in the mid-1930s, and focusing on young Francie Coffin, Meriwether has chosen a time similar to the time of publication—a time when a dream had in many ways been shattered by events but when the older generation, with memories of better times, could hold on to hope that better conditions would come again. Alice Childress's *A Short Walk* (1979) and Rosa Guy's *A Measure of Time* (1983), sweeping through the years and spanning the history of Harlem from its heyday through its decline, both conclude with early episodes in the civil rights movement, narratively connecting the two periods.

In choosing Harlem as the setting of these novels, Meriwether, Childress, and Guy have chosen the world they know best. All were raised there, were the children of parents who came there during its heyday, and personally experienced the years of its decline. All have aligned with the more militant end of the activist spectrum and have written novels that insist on the interaction of public and private lives, as well as the connections between seemingly contradictory strains of the long walk for racial justice. These three novels set in the Harlem Renaissance and its immediate aftermath leave readers with uncertainty about the future of the characters, and by exploring how the civil rights movement relates to the Harlem Renaissance, invite consideration of what might be the next stage—in the wake of the modern civil rights movement.

Of the characters in *The Color Purple* (1982), only Nettie, Samuel, and Corrine travel to the North, and the only account of that experience comes from Nettie, who writes to Celie about Harlem as if it were paradise: "New York is a *beautiful* city. And colored own a whole section of it, called Harlem. There are colored people in more fancy motor cars than I thought existed, and living in houses that are finer than any white person's house down home. . . . They live in such beauty and dignity. . . . They were all dressed so beautifully, too. . . . And all the people have indoor toilets, Celie. And gas or electric lights!" (114). Nettie apparently never considers that behind the glitter and polish there is an underclass struggling to survive or that the fancy cars she admires may have been paid for with money acquired through prostitution or other illegal means. She seems unaware of those displaced Southerners like Grange Copeland in Walker's first novel, who travels to Harlem in 1926 only to join the hordes of vagrants who never find legitimate work. In sharp contrast to Nettie's idealized version of Harlem is the complex and contradictory Harlem of *Daddy Was a Number Runner*, *A Short Walk*, and *A Measure of Time*.

In varying degrees of complexity, these three Harlem novels project the stories of the private lives of their female protagonists into the public fray of competing worldviews that gave to Harlem its vitality and its peril. While Nettie seems to have encountered only the middle-class churchgoers, Francie Coffin, the protagonist of *Daddy*, the child of a number runner, meets petty gamblers, pimps, child molesters, and prostitutes in her neighborhood, as well as street-corner revolutionaries and famous evangelists. Cora Green in *A Short Walk* consorts with Marcus Garvey and other world changers and earns a living in the world of entertainers, gamblers, and bootleggers. Dorine Davis in *A Measure of Time*, who moves in and out of all these worlds, acquires friends and acquaintances ranging from the most degraded down-and-outs to the elite members of the Harlem intelligentsia.

Although some of the celebrities that created the Harlem Renaissance were not themselves raised in the South, Harlem of the twenties and thirties, with many of its citizens the children and grandchildren of slaves, was still limited by what Rosa Guy calls "the obstacles of a slave past."[7] Like those who headed North on the Underground Railway during the previous century, many had taken a stand for freedom and against accommodation to a racist society. Having left the isolated black communities of the South and made it to Harlem, they had taken a first step toward determining their own fates and participating in a larger community, but major characters in all three novels retain their ties to the South and continue to take responsibility for those they left behind. Adam Coffin, Francie's "Daddy," takes in a family of starving sharecroppers in the midst of the Depression when there is hardly enough food to sustain his own family. Francie overhears a neighbor defending his activism, explaining that he has "to care what happens to black people in Alabama," since "what happens to them down south is part of what happens to us here in Harlem" (96). In *A Short Walk*, referring to his involvement in fighting southern as well as northern racism, Cecil explains to Cora that there is "little room for love until we solve this race hell" (240). In *A Measure of Time*, Dorine, observing the riots following the "Scottsboro mess," discovers "that when a black feller got cut in Alabama, folks in Harlem bled" (199).

Not surprisingly, these novels contain similar scenes and characters. In both *Daddy* and *Walk*, for example, characters trying to cope with the Depression hold house-rent parties, lie to welfare workers in order to qualify for relief, create recipes to make the otherwise inedible surplus food palatable, and visit Father Divine's establishment for an abundant good meal. Both *Daddy* and *Measure* include scenes in which characters riot to protest the Scottsboro case, celebrate Joe Louis's victory over Max Baer, and suffer losses when big-time gangster Dutch Schultz takes over the numbers racket in Harlem. There are even more common elements and parallel scenes in *A Short Walk* and *A Measure of Time.* The protagonist of each spends her childhood and youth in the South, comes to Harlem in her late teens, supports herself through a variety of legitimate and illegitimate activities, and is used and abused by the men in her life. Cora and Dorine are refused service in the same Harlem restaurant and stage spontaneous sit-ins in the dining cars of Jim Crow trains. All three novels weave real historical events and characters into the narrative of personal lives, and each of the protagonists is affected by what Dorine calls "all the world-changing" (139). Francie regularly listens to Adam Clayton Powell, Jr., speak of lynchings and the need for blacks to free themselves from the bondage of racism (53); through her boyfriend, Cecil, Cora becomes acquainted with Marcus Garvey; Dorine encounters both fictional and historical world changers, including W. E. B. Du Bois.

In each of the three novels, characters are conscious of the interaction of private and public lives. *Daddy* focuses on the impact of public events on private lives, while *Measure* and *Short Walk* consider the subtle ways that private individuals affect history. While writing *A Measure of Time*, Rosa Guy has said, she came to believe that "the survival of black folks in the United States was due mostly to the great effort of black women, the Doreen Davises."[8]

Narrated by the female protagonist speaking in her own voice, these novels convey the value and significance of the Harlem experience through one who experienced it first-hand, rather than from the distanced perspective of the historian. But there are significant differences in the narrative strategies used to connect that lived experience to a later time. In *Daddy*, young Francie Coffin tells her story in her own adolescent voice with no temporal distancing; she is speaking from 1935 as she was at that time. The narration of *A Short Walk* alternates between a third-person omniscient narrator with a "once-upon-a-time" perspective and Cora's spontaneous voice speaking in present tense. Dorine Davis in *A Measure of Time*, like Claudia MacTeer in *The Bluest Eye*, only occasionally looks back from a point in time removed from the events she relates to compare her experiences with those of some later and very different time. In a contemplative voice, she evaluates the past from a time that is clearly after the civil rights movement. But for most of the novel, Dorine narrates with immediacy that renders her story vivid and present, and readers are not so much looking back as they are swept back to the past.

By 1970, when the first of these three novels was published, Harlem had changed enough to be equated in most people's minds with the worst aspects of ghetto life—crime, riots, poverty, and drugs. No longer the mecca of black professionals, intellectuals, and entertainers, Harlem, once the place to escape to, had become for many a place to flee from. What was gone from Harlem in 1970, 1979, and 1983 was in a sense gone from the political life of the nation, now bereft of powerful leaders speaking on street corners, in town halls, and on the campaign trail. Gone was the New Deal, the New Frontier, and the Great Society; in their place were the Vietnam War, Watergate, and a new generation of urban poor. Gone too, were W. E. B. Du Bois, Marcus Garvey, Malcolm X, and Martin Luther King, Jr. While *The Bluest Eye* indicts a society that destroys individuals by exclusion and *The Color Purple* celebrates the private world of family and friends, these novels, even while they expose the flaws and failures of Harlem at its best, affirm the sense of community—of belonging to something bigger than themselves—that made it possible for individuals to hope for and work for a better future. However despondent young Francie Coffin might be at the end of *Daddy*, readers know her as a resilient and tough little girl, given to dark moods which she overcomes. We know that her mother has continued to nurture hope that one day they are "gonna move off these mean streets" (158) and that Francie, who is destined to go to college, will never "have to do no domestic work for nobody" (173).

The one public figure who plays a role in all three novels is Marcus Garvey: in *Daddy*, set in the mid-1930s, some ten years after Garvey's downfall, street speakers still cite his teachings to support their own stands against racism; *A Short Walk* relates Garvey's heyday and his downfall, while demonstrating the hold he continues to have on Cecil and presumably other followers; in *A Measure of Time*, even his enemies, Du Bois and Frazer, are working to save him from deportation. Like the heroines of these novels, whose lives were marred by private failures, Marcus Garvey, who failed in such a grandiose way in the public arena, left a legacy that made future progress possible. Garvey demonstrated that large numbers of African Americans were ready to follow a leader who acclaimed black dignity and pride, and the spirit and fire that he ignited may well have persisted in the minds of his followers long after he was gone, thus paving the way for the civil rights movement.

The women of Garvey's generation helped create the future in different ways. Mrs. Coffin in *Daddy*, Cora in *A Short Walk*, and Dorine in *A Measure of Time* lived what to many would seem failed lives. Abandoned by the men they loved, tainted by their connection with the underworld, and doomed to struggle alone to support themselves and their children, they nevertheless provided the resources for their children not only to succeed in mainstream America, but to join the new generation of world changers.

There is a sense in which the term *Harlem Renaissance* is a misnomer, in that the period in the late teens, twenties, and early thirties that carries the name was in fact a first widespread flowering—or birth—of the creative impulses, artistic and literary talents, and intellectual endeavors of African Americans. If that is the case, then the works of Meriwether, Childress, and Guy may be seen as part of a true renaissance, or rebirth, of Harlem's creative spirit nurtured by the Harlem Writers Guild that Rosa Guy helped to found in the 1950s.

Daddy Was a Number Runner

Louise Meriwether's *Daddy Was a Number Runner* is a historical novel in the sense that its fictional characters, who interact with and refer to historical figures, are presented as living in the ambience of a particular place and time. But beyond that, the thematic thrust of the novel is tied to the protagonist's determined effort to acquire historical sensibility. Francie Coffin may serve as a model for young readers attempting to understand the relationship between historical reality and personal identity. *Daddy* is similar in a number of ways to *The Bluest Eye*, published in the same year.

Each novel employs a first-person narrator whose progress is contrasted with the decline of a childhood girl friend. Set in cities, both focus on girls approaching puberty, whose parents are struggling with poverty. The action of *The Bluest Eye* extends from fall 1940 to fall 1941, and the events in *Daddy* take place from June 1934 to fall 1935. Unlike the characters of *The Bluest Eye*, however, those in *Daddy* are intensely conscious of the importance of public events in their lives and of the role of the community in their daily survival. Missing from the Breedloves' experience is any kind of supportive community; those who survive in *Daddy* do so in part because of community. Like Pecola Breedlove, Francie thinks of herself as ugly and undesirable, but Meriwether's novel is not primarily about this essentially private and social concern. Unlike Pecola, Francie does not dwell on her personal inadequacies; in moving outside the self, she gradually acquires what she needs to survive in the threatening environment of Harlem in the mid-1930s. While Pecola is obsessed with whiteness and the painfully naive notion that blue eyes would somehow assure her happiness, Francie, who is fortunate in being intelligent and having parents who instruct her in the ways of the world, struggles to understand the society she lives in and to find ways out of poverty.

Questions about the individual's social responsibility come out in *The Bluest Eye* as Claudia blames herself for Pecola's downfall, but Morrison's novel does not suggest the social or political mechanisms by which Pecola's lot might have been improved. *Daddy*, on the other hand, includes a number of characters whose political agendas address the problems of racism, poverty, and sexual exploitation, and the twelve-year-old protagonist transcends her own personal pain by concentrating on others—her brothers, her father, her friend Sukie—and on the public world in which her own private domain is so precariously set. Pecola, absorbed by the personal, is eventually permanently imprisoned in her own private world. Morrison's story of a family tragedy and the irreparable damage inflicted on a helpless child implies without specifying the mechanisms, the necessity of public and political intervention to counter the long history of destructive public policies.

The Bluest Eye dramatizes the tragedy of growing up in a family that equates blackness with ugliness; *Daddy* affirms the black-is-beautiful movement at its peak in 1970, by including characters who appreciate their own beauty. While no one counters Pecola's perception that she is hopelessly ugly, Francie's father opposes her perception that she is "skinny and black and bad looking" (15). A "giant of a man . . . dark brown, black really, with thick crinkly hair and wide laughing beautiful mouth" (21), he admires women whose features do not conform to the white aesthetic: "Daddy was always saying somebody was beautiful—some black girl with thick lips and a wide nose" (37). While Cholly Breedlove conveys to his family his shame and self-hatred, Adam Coffin inspires his children by repeatedly telling the mythic tales of their great-great-grandmother Yoruba, "the only daughter of Danakil, the tribal king of Madagascar" (75). Although no one in the family particularly believes the stories, they contribute to Francie's sense that she has "a past to be proud of" (77). Toward the end of the novel, when Francie discovers that her father is living with another woman, she summons Yoruba to accuse him: "You forgot about Yoruba, Daddy. You forgot you was one of Yoruba's children" (164).

Narrating her story in her own young adolescent voice, Francie moves steadily through time, recording her journey from naive childhood to knowing youth. More than *The Bluest Eye*, *Daddy* is addressed to a youthful audience, but like *The Bluest Eye*, the thrust of the chronological narrative contributes to the novel's affirmation that progress is possible even in impoverished communities in depressed times for those who learn to read the times and act accordingly. Particularly attuned to the interaction of private and public events, Francie recalls the Scottsboro riots, the shooting of Dutch Schultz, and the Joe Louis/Max Baer fight at Madison Square Garden; she matures as she learns to understand how the public arena informs private lives. As the child of a number runner whose survival depends on the whims of the police and the fate of gangsters like Dutch Schultz, Francie learns early that public events influence her private world.[9] When racketeers gun Schultz down, private and public history coalesce since "old Dutch . . . was head of the numbers" (185)—her father's racket. Unlike many of her counterparts, Francie reads newspapers, listens to street speakers, and attends the political sermons of Adam Clayton Powell, Jr.[10] She recounts events that will lead to World War II, particularly Mussolini's invasion of Ethiopia.

Clearly Meriwether intended Francie's story to alert young readers to the dangers of black ghetto life and to the behaviors and attitudes that will protect them from those dangers. In a voice that fuses innocence with street smarts,[11] Francie reveals how, in spite of living in a violent, racist, impoverished community, she learns to depend on herself and to resist the dangerous temptations that surround her.[12] With the guidance of flawed but loving parents, she develops "historical sensibility," a quality that Meriwether believes is powerful in bolstering "a people's struggle for power and fulfillment."[13] She learns where she came from, who helped, and where she might go; she escapes the perils around her through awareness of reality rather than by clinging to the romantic notions that torture Pauline Breedlove and her daughter, Pecola.

Like Lorain, Ohio, in the last years of the Depression, Harlem had its casualties, and in the course of a year, Francie, like Claudia MacTeer in *The Bluest Eye*, becomes aware of the destruction of lives that takes place around her every day. When the novel opens, on June 2, 1934, twelve-year-old Francie lives with her mother, father, and two brothers, James Junior and Sterling, in a vermin-infested apartment in Harlem. As a number runner, her father is trusted and respected by the neighbors, who are also his clients. Mrs. Coffin and the children help with the business, collecting money and hiding the evidence of their activities from the police. Compared to that of many of her friends and neighbors, Francie has a happy family. When the novel opens, her parents, who live together, are sober, considerate of one another, and conscientious about caring for their children. But the family's stability is vulnerable to violence, racism, unemployment, and poverty.

The first serious threat comes to the Coffin household when James Junior arrives home late after being initiated into a street gang called the Ebony Earls. Mr. Coffin punishes his son with a beating while Francie cries and listens to the blows. Such discipline is the only domestic violence that Francie knows, but in the streets of Harlem, she encounters violence at every turn. Her best friend, Sukie, responds to her own unhappy home by fighting with Francie, leaving her bloody and bruised. Sonny, another playmate, threatens to throw Francie across the alley from one roof to another and, to prove his meanness, kills his grandmother's pet cat by dropping it off the roof. As the year progresses the stakes are raised: three of the Ebony Earls are sentenced to death for murder, and Sukie's sister, China Doll, stabs and kills her pimp. The violence that afflicts Francie's small personal world is magnified many times in the society at large, where Francie continually encounters violence and senseless death: a white policeman picks a benign drunk off the street and beats him on the head and shoulders (55); riots follow the rumor that police shot a black child caught for shoplifting (136); police shoot and kill youths watching the riot (138).

More subtle than the constant threat of violence is the degrading racism and sexual harassment that Francie encounters each day. A white girl does not invite Francie into her house when she goes to visit; another refuses to touch her. Her white sewing teacher dismisses her desire to be a secretary, suggesting that she learn domestic skills if she expects to get a job. Teaching typing and shorthand to Negroes, her teacher insists, only creates false hopes and frustration (132). Throughout the novel, young Francie is sexually harassed by older white men and young black boys. On the stairways of tenement buildings, in dark movie houses, at the bakery and the butcher shop, and in public parks, men accost her, offer her money or food, and demand sexual favors. Most of the time she resists, but occasionally the prospect of a dime or an extra soup bone tempts her to submit to their clumsy fondling. When Sonny, one of the neighborhood boys, tries to rape her, Francie fights him and escapes. Watching her friend Sukie drift into prostitution, Francie soon determines to avoid such a fate.

The threatening sexuality of Francie's private world pervades the larger society, which, according to Sukie, provides black women with only five choices: "Either you was a whore like China Doll or you worked in a laundry or did day's work or ran poker games or had a baby every year" (187). Undaunted by racism and determined to succeed, Francie rejects each of the options, and clinging to her mother's prediction that they will leave Harlem "one of these days" (188), she learns to be open to other possibilities. She learns because she is surrounded by adults who, in spite of their flaws, teach her what she needs to know to survive; she learns because she listens, observes, and reads.

Through regular attendance at the Abyssinian Baptist Church, Francie hears Adam Clayton Powell, Jr., talk about lynchings in the South and racism in Europe (53); she hears her father explain Powell's idea that people riot not because they are evil but because of unemployment, discrimination, and insult (137). The public outcry over the Scottsboro case teaches her that some people feel responsible for doing something about the racism that extends far beyond the boundaries of her school and neighborhood. When his wife accuses him of neglecting his children, Robert, a neighbor and member of the Black League for Freedom, loudly defends his activism so that all the neighbors—including Francie—can hear: "I got to care what happens to black people in Alabama. Nine colored boys are condemned to die because two white sluts said they raped them. Ain't that a bitch? Can't you understand that what happens to them down south is part of what happens to us in Harlem?" (96). In addition to reading newspapers, attending Powell's church, and listening to street speakers, Francie also reads novels. From Claude McKay's *Home to Harlem*, she learns about her predecessors in these "same raggedy streets" (101); *Uncle Tom's Cabin* makes her "feel black and evil toward whites" (105).

Unlike *The Color Purple*, which never mentions specific deprivations during the Depression, *Daddy* dramatizes the daily struggle that all but the most privileged and fortunate endured during that time. When even whites are unemployed, Francie's father has little hope of finding legitimate work; at the beginning of the novel the family's income is limited to an occasional "hit" in the number game and the tips he receives from running the numbers and playing the piano at rent parties. Even though his income is inadequate, Coffin refuses to apply for relief and opposes his wife's going out to work, insisting that she should be at home to care for her children.

And when times get worse and Mrs. Coffin takes a job as a domestic, at first part-time and then for six full days a week, the children do suffer. The family finally goes on relief and endures the absurd regulations of the welfare system, which require that they liquidate all savings, including the children's college fund, leaving nothing to build on when times get better. Although the relief check is not enough to feed the family, Coffin must hide any part-time work he does or risk losing it. On the rare occasions when he makes a hit and there is enough money for good food and new clothes for the children, family life improves, but poverty and the loss of personal dignity contribute to the disintegration of the family. In the course of the novel, James Junior leaves his teen-age gang behind and becomes a part of the adult underworld; Mr. Coffin, ashamed of his inability to support his wife and children, abandons them; Mrs. Coffin works six days a week; and Sterling, with an aptitude for science, quits school to become an undertaker because he does not expect success in a world where firms "hardly hire a black janitor," let alone black chemists (153).

Daddy places the characters' personal suffering in the broadest possible social context and demonstrates that the unemployment and poverty that afflict the Coffin family in 1934–35 are widespread in the society and have consequences that are everywhere apparent: Adam Clayton Powell, Jr., feeds a thousand a week in his free food kitchen (51); hungry people crowd into Father Divine's place (87); a sharecropper's family from Virginia is literally starving on the streets (118); a six-month-old baby dies in Harlem Hospital after being bitten by a rat (139); and street vendors huddle in the cold dressed in rags (142). For Harlem residents, public relief is far from adequate, and dealing with the social workers often requires putting on an act. One character tells Mrs. Coffin that "President Roosevelt said that money was to keep poor folks from starving. . . . Just go down there and act bad and they'll put you back on relief just to get rid of you" (121).

Both of her parents teach Francie to prepare herself to live in a better world than Harlem in the middle of the Depression. Mr. Coffin advises his children: "You better prepare yourself for the future. . . . Times gonna get better and you ain't gonna be ready" (77). After Francie's first and last day as a domestic worker, her mother tells her: "You don't have to do no domestic work for nobody. . . . You finish school and go on to college" (173). Even China, the prostitute, tells Sukie and Francie that they will have "a better break than she had" and that they had "better get ready for it" (109). Being ready means staying out of trouble and getting an education.

When Francie gets caught up in a riot, she sees the senselessness of people destroying their own neighborhoods, and when she reads about events that she has observed in the newspaper, she notes the discrepancy between reported news and reality. In a striking way, Francie discovers that she cannot always believe what she reads and that she must depend on her own investigations to know what is going on. At the end of the novel, then, Francie has learned to think for herself and to question what she reads and what others tell her. She is much less vulnerable than she was a year before.

In the summer of 1935, "the strangest thing" happens to Francie. Watching a Ken Maynard cowboy movie, she suddenly realizes that she is rooting for the Indians. By opposing both racism and exploitation in the context of the movie, she prepares herself to resist those who have been exploiting her. First comes Max the Baker who offers her cookies in exchange for a secret fondle, but Francie gives him a quick kick in the groin and runs off determined to do the same thing to "that lousy butcher" the next day (158). When Sukie suggests they go to the park where men offer them a nickel for a peek under their skirts, Francie says no again (159).

As Francie develops "historical sensibility," she locates her own life in its slave and African past and develops a sense that she is caught up in something bigger than herself: "We were all mixed up in something together, us colored up here in the north, something I couldn't quite figure out" (123). Acquiring self-knowledge in *Daddy* and in other novels treated in this book does not mean exploring the inner workings of the autonomous psyche—but rather setting the self in the context of family, neighborhood, ethnic group, and finally the larger society. From Adam Clayton Powell, Jr., and from street speakers like her neighbor Robert, she learns that history is change and that at least some people plan to be involved in making history: by "boycotting the stores on 125th Street" (166), getting "a better education," and building "Negro economic and political freedom." Although she does not yet know what she can do, Francie publicly condemns those who refuse to "do something" to help themselves and change the world (169).

The temporal gap between this novel's setting in 1934–35 and its publication in 1970 was, of course, the time of planning for and implementing the ventures that made up the civil rights movement. Most people in 1935 were preoccupied with survival, but some, like Francie's parents, struggled to teach their children to be ready when "times . . . get better" (77). By 1970, when the novel was published, some people had escaped the ghetto and joined the struggle for civil rights. A few, like James Baldwin and Claude Brown, had written books about their escape. Many more, however, were as trapped in the ghetto in 1970 as the Ebony Earls were in 1935. As a historical text, *Daddy* suggests the parallels between that time and its own and warns about the dangers threatening young lives then as now. But as the narrative unfolds, it highlights the specific actions that facilitate social change: one generation making sacrifices for the next, youths committing themselves to education, and individuals joining in the political process.

In the final scene, sitting on a stoop with Sukie and Sterling, Francie counters her despair by recalling her mother's prediction that they will some day move away from Harlem. Sterling grunts and says "shit," and Francie echoes him. At this point, the novel seems to end on a note of despair. If it had been published in 1935, it would have been a despairing book: readers then would not have had much confidence that conditions would improve. But readers in 1970 and after knew that times did change: World War II began; the Depression ended; Roosevelt created the Commission on Fair Employment Practices. And there is every reason to believe that Francie's life did improve with the times—she is, after all, getting an education and preparing for the future. By 1970, the process that A. Philip Randolph had begun in 1941 by threatening a massive march on Washington had played itself out. Francie would have been thirty-two years old at the time of *Brown v. Board of Education* in 1954 and forty-three at the time of the Voting Rights Act in 1965. She might have become an activist like Rosa Parks or, as the fact that she is writing her own narrative suggests, a writer—perhaps a Louise Meriwether, whose own father was a number runner in Harlem during the Depression.

In 1970, many of the first readers of this novel, however, may have looked at the world around them and concluded with Francie that the prospects for a better world were best summed up in a four-letter Anglo-Saxon word. By the end of the summer of 1968, Medgar Evers, Malcolm X, Bobby Kennedy, and Martin Luther King, Jr., had all been assassinated; many American cities had been ravaged by riots; violence and bloodshed had dominated the streets of Chicago during the Democratic Convention. In 1970, the United States invaded Cambodia; students were killed by National Guardsmen at Kent State and by police at Jackson State; and Richard Nixon publicly referred to dissident students as "bums." What prevents Francie's four-letter summary of the State of the Union from being as appropriate for black Americans in 1970 as it was in 1935 is the history of the thirty-five years intervening between the novel's setting and its publication. Though the five years during which the novel was composed and prepared for publication may have seemed like backsliding, the long view from the middle of the Depression is still mainly one of progress.

Daddy Was a Number Runner invites various responses from readers. To resourceful young blacks already enjoying the benefits of the civil rights movement it speaks of what life was like for their parents and grandparents who paved the way for them. To those still trapped in ghetto poverty it speaks of education as a way to a better future; to others in 1970 and beyond, its protagonist's development of historical sensibility suggests looking around them to the politics of their own times.

But there are other lessons to be learned from *Daddy*. Robert, the political activist who speaks on street corners, may speak as well to those newly attracted to the black power movement:

> I tell you, brothers and sisters, the black man in this country must take his own life. The crying Negro must die. The cringing Negro must die. If he don't kill hisself the environment will, and we been dying for too long. The man who gets the power is the man who develops his own strength. I ain't talking about strength in his muscles but in his mind. We got to get a better education. We got to build Negro economic and political freedom. And if we don't, in fifty years from now, or sooner, this country will be bloody with race wars. (169)

Urban readers in 1970 and beyond would know that Robert's predictions are correct. The country was bloodied by race wars, and sooner rather than later.

But there were also changes for the better: black Americans had achieved at least some economic and political freedom, and many had improved educational opportunities. Such changes had occurred in part as a consequence of historical forces external to African-American culture—World War II and the end of the Depression—but also as the result of the efforts of early civil rights leaders, such as Adam Clayton Powell, Jr., A. Philip Randolph, Ella Baker, James Farmer, and those like Francie who were influenced by them. The long struggle for civil rights was generated during the years that Francie was growing up, and it began to gain momentum about the time she would have reached adulthood. *Daddy* reminds others who have escaped Harlem and its equivalents all over the country how much they owe to those who made it possible.

At the end of the novel, the men who shot Dutch Schultz have been released while three of the Ebony Earls who were tortured into a confession have been sentenced to death for the murder of a white man. Though she "don't wanna be no whore," Sukie has already had sex with her sister's pimp. Neither Daddy Coffin nor James Junior comes around anymore. Francie, on the other hand, has taken a different path: she has rejected racism, exploitive sex, and violence; she reads newspapers and good books and does well in school. And she has a mother who is willing to work extremely hard to make a way for her daughter. Though neither Francie nor her mother sees a way out, they are working to create that way. Readers who did not see a way forward in 1970 may have taken heart.

Daddy might be viewed as a work about survival. But then of course, so was *The Bluest Eye*. In its focus on Claudia MacTeer, one who has succeeded by deserting the Pecolas of her world, *The Bluest Eye* is a survivor's story too. Claudia, the survivor, looks back with sadness at one she left behind in a metaphorical prison—if not concentration camp—of the mind. *Daddy*, addressed to survivors and potential survivors, focuses on one who is determined to make it, who has identified her good with that of the community, and who acknowledges education and political action as appropriate tools for creating a better future.

A Short Walk

While *Daddy*, with its setting limited to a few blocks in Harlem, recounts the events of the life of young Francie Coffin in a single crucial year in the mid-1930s, Alice Childress's *A Short Walk* relates the major events of Cora Johnson's life, beginning with her birth in 1900 and ending with her death almost five decades later. A comprehensive novel that moves from South to North to in-between, *Walk* is Dickensian in its sweep across the social spectrum and its scanning of the effects of social policies on the lives of individuals struggling to meet their needs in a society structured to exploit them. *Daddy*, published in 1970 just as the younger generation was first enjoying the benefits of the movement, ends with the young protagonist preparing to begin her life. *A Short Walk* was published nine years later, when it was clear that the more ambitious goals of the movement were still unfulfilled. It ends with the death of the protagonist, whose life has paralleled and occasionally intersected the first wave of activism leading to the civil rights movement.

The first ten of the twenty-nine chapters of *A Short Walk* focus on life in Charleston during the first two decades of the twentieth century, and except for a few episodes that treat the experiences of Cora Johnson traveling in the South or accompanying her boyfriend Cecil Green on his adventures with Marcus Garvey, the remaining chapters are set in Harlem. Like *Daddy*, *A Short Walk* explores the impingement of public events on private lives, but its temporal scope allows for a more complex and sustained examination of individuals struggling with the conflicting demands of private need and public commitment.

Contributing to the novel's complexity is its alternating point of view. The periodic shifts from third-person to first-person narration evoke a large public panorama from which Cora's highly personal, colloquial voice issues. The novel's two voices sometimes reinforce each other and at other times stand in ironic contrast. Chapter 5 begins in Cora's fresh adolescent voice, reporting and interpreting what she has been told about why World War I is affecting their family's life. The third-person narrator then confirms Cora's perception that their intensified poverty is the consequence of the influx of immigrants who are taking the jobs previously held by blacks. Sometimes, however, Cora's imagined version of the past is different from that reported by an omniscient and presumably reliable narrator. For example, when she learns that her father was white, she concludes that he was a "lowdown thing," but the narrative account of her mother and father's brief affair—abruptly broken up by his parents—presents him as sensitive, very young, and quite smitten by Cora's fifteen-year-old mother, Murdell. Though some of the third-person chapters are mainly dialogue with little narrative comment, that objective voice periodically intervenes in the dramatic presentation of events to explain what motivates the characters. Cora reports that Cecil cannot even hear her when she twice tells him that she is pregnant with his child, while the narrator's less sympathetic view of Cecil suggests he may have ignored that unwanted news.

Shortly after Cora is born, old Mr. July, a former slave, stands over the body of Cora's dead mother and predicts a future of progress and a better life for Cora: "Most forty years done pass since the last day a bondage, so Murdell's baby gonna someday walk where we now can't go, live to say what we can't, gonna taste the sweet years to come. Her life will live easy" (16). From the point of view of the late 1970s, Cora's "short walk" through life seems anything but easy, but from the perspective of a former slave who earns his living collecting dog droppings to sell for fertilizer, Cora does go where he could never go and speaks what he could never say.

The measure of progress in this novel depends on the historical perspective from which it is viewed. In terms of actual progress, the walk from slavery through the period of World War II was a short one; calculated in the months, years, decades of the "learn to wait" and "bide your time" that Cora hears from her adoptive mother (72–73), the walk has been very long indeed. From the perspective of civil rights activists of Cora's generation, the journey seemed by 1979 to stretch endlessly and to go over the same ground again and again. For Cora, however, the walk was taken in fits and starts and never with the intention of staying constantly on the road. The degree of her involvement in protest and social movements varies under the influence of external events and the demands of her private life, but from early childhood, she takes an active though impulsive role in opposing racism.

Her loving adoptive parents, Bill and Etta James, teach Cora about the problems of living in a racist world, but they give her contradictory instructions about how to deal with racism in everyday life: Etta counsels resignation and patience; Bill advocates resistance and clings to hope for a "better future for the Negro race" (85). Cora soon learns that public policy affects the quality of her life, and she begins early to fulfill Mr. July's prophesy by rebelling against the Jim Crow world into which she was born. Walking past a "Whites Only" public park on her fifth birthday, she convinces Bill that they should defy the law and enter the park; when a white policeman runs them out, Bill gives her a history lesson, explaining about lynching, white supremacy, and slavery.

That afternoon at the minstrel show, Cora notices that the audience is segregated according to race and class, with blacks sitting separate from poor whites, and well-to-do whites sitting behind and above them all. In spite of the pervasive racism of the blackface show, Cora enjoys the humor, but the high point of the day comes when an attractive woman sings about freedom from oppression and slavery with an earnestness and beauty that is inconsistent with the crude humor of the rest of the entertainment. The whites react, disrupting the show with threats of violence. As she watches blacks and poor whites hurry "down separate but equal dirt roads" (44), Cora learns "about the colored and white of life" (46).

Growing up in Charleston, Cora becomes increasingly curious about what it means to be black in a white world. Listening to her parents and teachers talk, she learns that blacks blame themselves for what goes wrong and that raising questions about race is discouraged. Reading from a textbook she asks if Ned, the boy "with a cart and a pony," is white (53), only to be ridiculed by her black teachers. After whites object to the "political nature" (64) of a church pageant and threaten to invade the performance, the congregation capitulates rather than risk trouble, but young Cora argues that "Times oughta change" (70). Her father urges her to hold on to her outrage about Jim Crow, insisting that to tolerate discrimination is to give white people what they want. When her mother explains that people have been lynched for thinking like that, Cora suggests that there may be "a time to die for whatcha believe" (71). Though she sometimes gives in to her mother's admonitions—"Learn to wait . . . and just bide your time. Your day will come" (72–73)—Cora periodically defies Jim Crow.

Cora is an adolescent when she begins to see that people's behavior is conditioned, if not determined, by public realities. After a neighbor brutally murders his wife, she wonders whether he might have been motivated by "the price of oranges, or the colored and white signs . . . or some other such thing" (91). Not long after she comes to believe that her own survival may depend on understanding public attitudes and social conditions. On the day her father dies—in a scene similar to the interruption of Cholly Breedlove's first sexual encounter in *The Bluest Eye*—three white boys accost Cecil and Cora as they are about to make love. They humiliate him and threaten to rape her. Unlike Cholly, however, Cecil fights back, and even though he has "done half kill a white boy" (111), Cora wants to call the police. Cecil's Aunt Looli, a prostitute, rescues them and teaches Cora that, in a racist world, she cannot count on the police, and that she must "grow up and catch on to how to care for yourself" (112).

After her father's death and her boyfriend Cecil's departure for New York, Cora, at her mother's urging, marries a man she does not love for the security he appears to offer. But when he turns out to be a sadist and wife beater, Cora, still in her teens, escapes to Harlem. Her father's cousin Estelle introduces her to life there and teaches her how to survive in the city: how to hold parties to raise rent money, sell illegal whiskey, and run a "respectable" boarding house. In and out of cousin Estelle's establishment move a diverse group of sleeping-car porters, singers, dancers, and the lively and good-natured Napoleon Ramsey (Nappy), who teaches Cora to deal cards, to sing and dance, and to play straight-girl for his comedy number. With help from Estelle and Nappy, Cora supports herself in the worlds of gambling and entertainment.

She soon discovers, however, that leaving the South does not mean leaving racism behind. On the trip to Harlem, she rides the Jim Crow coach, "the first one behind the engine," not just through the Southern states but all the way to New York. Estelle explains that the North "is just as mean as down home, only with no colored and white signs" (142). Cora learns that segregation is maintained in New York by such devices as seating blacks in Broadway theaters "in the same row with other coloreds" and seating them but refusing to serve them in restaurants (156). Outraged by such treatment, Cora can never completely go along with Nappy's insistence that blacks "have to first make a livin and take care a the race problem later" (157). While Nappy chooses to focus on what immediately impinges on his life and work, Cora often protests the ubiquitous racial discrimination she encounters.

After some two years in Harlem, Cora runs into her first love, Cecil Green. Within hours of their reunion Cecil begins to teach her about another side of Harlem life: the political and public arena presided over by the dynamic Marcus Garvey.[14] An ardent follower of Garvey and supporter of the Universal Negro Improvement Association (UNIA), Cecil spends his time studying, lecturing, and writing for the UNIA's weekly newspaper, *The Negro World.* He teaches Cora about "the poor of the West Indies, the history of Afro-Americans, The African Orthodox Church, the plight of South American cane cutters" (165), and about Garvey's determination to free all blacks from domination by whites. Cora agrees to pose as Cecil's wife and accompany him on the first voyage of Garvey's Black Star Line—"I must go with the Black Star Line—I need it for my soul" (183)—but as it turns out, Cecil does not expect her to take a public role in the movement. During the ill-fated trip to Cuba and Central America, Cora sometimes enjoys herself, eating fine dinners, drinking champagne, riding in limousines and dancing in the presidential palace in Havana. But more often, Cecil leaves her to her own devices.

When the trip is over, Cora tells Cecil that she is pregnant, but he continues to talk about Garvey. She tells him she loves him, and he responds by offering to get her a cab. Absorbed in his public life and convinced that the UNIA "will triumph in the end," Cecil warns Cora that "there's little room for love until we solve this race hell" (239–40). Though she continues to be enthusiastic about social reform and periodically engages in her own private, usually impulsive protests, Cora cannot accept Garvey's dismissal of all whites or his willingness to collaborate with the Klan in advocating separatism. Cora not only has white friends, but she recognizes that racism is not limited to whites and that dark-skinned women have "a harder row to hoe" than those whose skin is "tantalizin tan" or "sun-kissed bronze" (214). Her rejection of Garvey's principles and Cecil's total involvement in politics leads Cora to decide that either she is "walking away from the Black Nationalist Movement," or "it's walking away" from her (222).

Presented with Cecil's view that solutions to the "race hell" must come before love and Nappy's insistence that immediate self-interest comes before solving the "race problem" (157), Cora sees no way to resolve the conflicts between the demands of private and public worlds. She finds herself shifting from one to the other in response to changing circumstances. From the moment she and Cecil return from the trip with Garvey, Cora realizes that her personal needs are in conflict with Cecil's public commitments. Even fatherhood does not move him toward private life. For Cora, however, pregnancy means that she must shift again from rebel to survivor; when she considers her options and thinks about how she will take care of a baby, she quickly concludes that "money is the answer" (228). And money means going on the road as an entertainer with Nappy and enduring the Jim Crow world, which she temporarily learns to do with good humor. Willing to pretend that the baby is his and to care for Cora during her pregnancy, Nappy nevertheless expects "a soft bosom to lay his head on comes night" (241).

After the baby is born, the novel skips quickly through the last of the twenties into the early years of the Depression. Cora becomes the manager of her own traveling show and succeeds in making it pay, "depression or no" (256). Free, at least for a while, from dependency on men, she once again shifts from the personal to the public arena when she finally rebels against the Jim Crow practices she encounters on the road. On a long "backbreaking ride to Georgia" (256), Cora stages a spontaneous sit-in in the dining car; by demanding service and threatening to turn the "coach into Lindberg's [*sic*] airship" and running it "nonstop from here to Paris" (260), she gets what she wants. In Georgia, Cora demands that a theater manager open the "white only" dressing rooms to her company. When he refuses, Cora anticipates the language of Martin Luther King, Jr., and other moderate leaders of the civil rights movement: "Call the police if you wanta. Lettum come. Throw me in jail, lynch me—I'm ready to die! This is my last stop on the Jim Crow line!" (270). Significantly, Cora's defiance of the separate and clearly unequal world of the black entertainer coincides with her decision to give up show business and stay home with her child.

Back in Harlem in the depths of the Depression, Cora prepares to do what is necessary to survive, confident that she will pull through: "A depression can break some people, but none that I know" (278). Sharing with friends, learning to beat the welfare system, holding a house-rent party, stopping by Father Divine's for a meal, Cora is ready when opportunities arise, and for her they always do. One Christmas toward the end of the novel, presumably about the time Clark was preparing to conduct the first doll test, Cora's "attitude about race pride" lands her "a lovely job" selling "colored dolls." Conducting her own informal doll test, Cora discovers that the younger children like the dark dolls, while those "who are old enough to be used to havin white dolls," hate them (297). Cora, unlike the mothers in *The Bluest Eye* who had no access to these dolls, subverts her daughter's infatuation with whiteness and buys her one of the "colored dolls" to go on her dresser top. Though she is probably too old to play with dolls, Delta gradually gets the message and accepts her identity as a black woman. Cora prepares to reinforce Delta's pride in being black by sending her to a private black college.

Throughout the Depression and into the forties, Cecil hangs "on to the battle of life with bulldog determination, afire with zeal for the ever-bright cause" (320), moving from the failed UNIA to the Harlem Labor Union. He speaks on street corners, addressing "those who may well chart the future . . . to carry out the beautiful rebellion" (321). His speeches are no longer about Garvey's Africa, but about black business, boycotts, and petitions (323).

Although neither Cecil nor Cora has succeeded in living a whole, balanced life—he has concentrated on changing the world, she on earning a living—their daughter Delta finds a way to do both. For years Cora has saved money for college, but Delta has other plans. She attends college only briefly before she returns home, determined to get her mother's permission to join the military, not so much because she wants to help fight the war, but because she believes that in the service she can begin "a perfect life filled with unlimited chance for careers, wealth, homes, and first-class travel by plane, train and cruising ships. . . . The first move: to break out of the restraint of Harlem and 'special' schools—to search for and find a fresh, new place" (322). But what Delta actually finds, contrary to the implied promises of recruiting officers, is a segregated military that discriminates against blacks. And so she finds her way to activism.

Much more than for her parents, history is on Delta's side. Coming of age during World War II, she becomes an activist at a time when organized efforts to change society are about to bear fruit. When the war is over, Cecil, in spite of his long-term dismissal of women from public life, is pleased that Delta is "protesting discrimination" in the military (329). President Truman's executive order desegregating the armed forces was the first significant step in desegregating American society, and as an activist within the armed forces agitating for change, Delta Garvie Green is one of the people responsible for that order.[15] Without consciously attempting to do so, Cecil and Cora have raised a child who is a pioneer in the civil rights movement.

Though *A Short Walk* is grounded in the history of the struggle for racial justice through the 1940s, interwoven in that context are issues of long-standing concern which came to the forefront in the 1970s, the most conspicuous being feminist issues. From the time that young Cora first longs to leave Charleston, South Carolina, by working as a cabin girl on a ship and learns that there are only cabin boys, she opposes sexual as well as racial discrimination. Cora has heard from her mother not only that she must "bide the time," but that "women must wait" (72). In spite of her determination not to wait for what she wants, Cora eventually recognizes that she has been waiting for a man to take care of her. It is on the trip with Cecil on the Black Star Line that she learns he has no money and that she will have to be her own "good provider" (203). But she concludes that providing for herself in a sexist society seems to require that she submit to the demands—sexual and otherwise—of the men who help her: "I can't get anywhere on this streetcar called 'the world' without having to pass some man who's always the conductor. And most of the time I have to pay my fare by layin out on the flat a my back, just to get him to take me wherever he might be goin" (241). Though she is angry about being exploited, she is willing to accept Nappy's kindness and even to go along with his pretense that the baby is his, and in spite of Cecil's neglect of her, she agrees to his request that she name their daughter for Garvey, even though he believed in keeping women in their place as servants to men.

From the objective narrative we learn that Cecil, a male supremacist, wishes that he had money so he could "put and keep her [Cora] in her proper place" (238). In a world where men rarely take care of women, it is never clear what Cecil's notion of that place might be. When he objects to his daughter's joining the military to fight white men's wars, Cora's delayed decision to give her consent seems motivated by her desire that her daughter have freedoms she never had: "My daughter can now be a cabin girl, in a manner of speakin, wherein I could not" (325). While Cora gradually gains clarity about her place in the world and does eventually provide for herself through a lucrative, though illegitimate business, Cecil continues to hold contradictory positions about women. In the end, his public position as a separatist and nationalist are in direct conflict with his personal life—he takes a girlfriend who is both white and a feminist.

In its walk through the twentieth century, the narrative thus comes full circle in its treatment of the relations between blacks and whites. Cora, after all, was born because her young parents defied Jim Crow. As children, both Cora and Cecil see whites as the enemy to be kept at bay, an attitude that prepares Cecil to hear Garvey's message. But soon after she comes to Harlem, Cora becomes close and lifelong friends with a white woman. Defending that friendship, Cora explains that she tries "to accept people just as they come wrapped" and treats "white folks accordin to how they act" (294). In the end, Cecil and Nappy have white lovers, and Cora is in business with a white man she admires. Outraged by the inconsistency between Cecil's nationalism and his willingness to take a white woman as a mistress, Cora finds, ironically, that she is the more thoroughgoing nationalist. Unable to have a physical relationship with a white man even though she is very fond of him, she says, "I don't preach on the street corner, but my nationalism is right here in this room" (312). In various ways, however, the main characters have moved closer to accommodation and cooperation, even relationship with whites; and Delta is working for integration, not separatist nationalism. But the novel does not finally endorse specific actions to bring about social change. Rather, it leaves the characters—and readers in 1979 and beyond—grappling with the still-unresolved issues of the black and white of life.

Although Marcus Garvey's movement failed in one sense, his effort to mobilize members of the black community to take part in a massive effort to change their status in the world paved the way for the civil rights movement. While Harlem of the mid-forties had declined a long way from its days of glory, members of a new generation were walking away from its decay, preparing to overcome the barriers that threatened their parents' public progress and private happiness. By 1979, a new generation had profited from the struggles of their elders, but they had lost the sense of unity and common cause that once fed the movement.

In the late 1970s, it was clear to separatists and integrationists alike that those committed to progress in eliminating racial discrimination would repeatedly retrace their steps and that the never-ending journey to the promised land of racial harmony would be fraught with conflict, frustrated by human shortcomings, and assaulted by history. *A Short Walk* speaks to those whose separatist alliances are challenged by personal, professional, and even political associations; and to those whose personal responsibilities seem to prohibit—or are used as an excuse to avoid—effective political action. It dramatizes the inherent contradictions in the political and the personal, revealing how the attraction to a woman—white, socialist, and feminist—is sufficient to override Cecil's avowed commitment to nationalist separatism; how Cora's rejection of separatism and her affectionate attachment to white friends does not leave her open to sexual relationships with whites.

In 1979, neither the nationalists (Garvey's heirs), nor the integrationists (King's followers), had seen their dreams realized. True integration, like willed separatism, was far from a reality; from the perspective of those still trapped in poverty and despair the need for a meaningful coalition of warring factions within the black community was as urgent as it was in 1945. On the night that Cora dies, Cecil has left her alone to attend a "*coalition meeting*" (330), presumably a meeting with various segments of the activist community but perhaps a rendezvous with his white girlfriend. Either way, the novel, in spite of its feminist subtext, does not finally denounce Cecil, but embraces him with the same sympathy that it offers the other characters. *A Short Walk* invites its readers to take the long view and to accept each others' inevitably conflict-ridden humanity, without giving up the struggle for human equality.

A Measure of Time

Like *A Short Walk*, Rosa Guy's *A Measure of Time* makes a panoramic sweep through the years of the Harlem Renaissance and its aftermath, using similar scenes, characters, and narrative techniques. Alice Childress's novel ends as characters are initiating the protests that will lead to desegregation of the armed forces; Guy takes her story ten years beyond to the early days of the Montgomery bus boycott. Like Cora in *A Short Walk*,

Dorine flees the South and a sexually abusive man when she is still in her teens and comes to Harlem in its heyday, where she soon becomes part of its many worlds. While still in the South, both women keep secret stashes of money which they use to escape, and once in Harlem, they do what it takes not just to survive, but to do so in style. Dorine, like Cora, maintains a long-term attachment to a man she does not marry but with whom she has a child, who at the end of the novel becomes an activist in the fight for civil rights.[16]

In spite of Cora's brief involvement with the Garvey movement and both women's public rebellion against Jim Crow in Harlem restaurants and in dining cars of trains, neither foresees playing a significant role in changing those aspects of society that both deplore. Like Cora, Dorine succeeds in earning money for herself and those who depend on her. The most significant distinction between the two narratives is the way the protagonists relate to the public events that are constantly shaping their private lives. While Cora is an active participant in the Garvey movement, Dorine's one flirtation with activism, inspired by the uproar of the Scottsboro case, is limited to a momentary impulse to join the Communist party. Cora attempts to understand and even occasionally to become part of the historical forces that govern her life, while Dorine responds to public events as personal insults. Though the man she loves has "nothin on his mind but some kinda freedom movement" (306), "Cora feels that sense of pride she takes in Cecil whenever she lets herself fully listen to him speak" (322). Yet, even though she sanctions Cecil's public commitment, she, like Dorine, deeply resents the personal price she has paid for her attachment to a public man. In spite of her resentment, however, Cora willingly contributes to Cecil's causes. Dorine, on the other hand, scoffs at plans to reform the racist South but unintentionally provides the economic base that enables her brother and son to join the civil rights movement.

A Measure of Time eulogizes Harlem during its golden age and laments its decline, while laying bare "the warped and twisted soul . . . of the southland" (247). Dorine's periodic trips to Alabama and her wanderings in and out of the various worlds of Harlem give the novel its basic structure and suggest that the migration to the North was a necessary precondition for the southern civil rights movement. Told in the first person, the narrative shifts from Dorine's younger voice, intent on the present and the private, to her older voice, which indirectly suggests the relevance of the public sphere. In the larger narrative context, both visions seem valid, though partial. Readers in the eighties concerned with racial justice were once again grappling with the best way to meet the competing demands of public and private life; like the characters in this novel, they arrived at different conclusions about whether to pursue personal ends, perhaps paving the way for another wave of activism, or whether it is better to engage directly in public action—whatever that might mean in the 1980s.

In the prologue to *A Measure of Time* Dorine recounts the important events of her young life: how she was raised in Montgomery by her grandmother; how she fled to Cleveland at the age of fourteen to get away from a white man; how that man left her an inheritance; and how she is traveling on a Jim Crow train to New York to join her boyfriend Sonny, having secretly left their baby with her sister.

Each of the novel's four books contains episodes set in Montgomery. In Book 1, Dorine goes back in time to relate that at the age of eight she worked in the home of a wealthy white man who repeatedly raped her. The rest of Book 1 focuses on Dorine's early days in Harlem and her turbulent relationship with Sonny, who introduces her to an exuberant gang of thieves.

In Book 2, Dorine joins the gang and acquires enough money to support her family in Montgomery and live the high life in Harlem. From May 1927 until the eve of the Depression, she mingles with famous entertainers, big-time hustlers, political activists, and writers. Totally preoccupied with the personal and frivolous aspects of her life, she is not impressed by her proximity to the great men and women of the day or by events that will radically change her world. On the day Lindbergh takes off for Paris, Dorine, wondering "Who the hell is Lindbergh?" recalls that she danced all night to Fats Waller's music, shouting "Lindbergh, Lindbergh, Lindbergh" and that her evening was spoiled when Sonny arrived escorting "the real Bessie Smith" (103–4). At "The Club" when a friend tries to introduce her to W. E. B. Du Bois and a man named Frazier (presumably E. Franklin Frazier, the social scientist) she is bored by the "talking, talking, all the world-changing" (139).

In the midst of this activity, Dorine is suddenly drawn back to Montgomery by news that her sister is ill, and on the trip home she attempts to use her expensive clothes and money to escape the indignities of the Jim Crow train, only to discover that they will not buy her respect—or a ticket on a sleeping car. When her own personal freedom is threatened by Jim Crow, Dorine argues for the need to get the "laws changed," but when a fellow traveler insists that there "ain't no law without power," she cannot imagine how blacks might seize power (117). She does, however, know something about the power that comes with money, and Dorine feels "guilty about not sending more money home" (145).

As Book 3 opens, the Depression is on, but Dorine prospers. She ignores the impact of public events on her private life until she is called south to care for her invalid brother-in-law. Avoiding the Jim Crow train, she travels by car, only to encounter segregated bathrooms, roadside restaurants that refuse to serve African Americans, and a Southern sheriff and "thin-faced redneck" who harass black motorists. In Montgomery, Dorine learns about the nine black youths arrested in Scottsboro for allegedly raping two white girls and about a gang of white men who respond to the news from Scottsboro by murdering an innocent black man. Hours after Dorine and a friend find his bullet-ridden body, they are back on the road heading north. Public events have entered her private world.

Scottsboro fever rages as crowds of blacks and whites gather in Harlem. Surprised that whites too are outraged and that "when a black feller got cut in Alabama, folks in Harlem bled" (199), Dorine is moved for the first time to take a public stand, but in a matter of days the impulse fades, and she goes back on the road with the gang to earn money to support the new man in her life and to send home to Montgomery.

Having sent her brother to medical school, Dorine goes south for his graduation, only to learn that he will not be allowed to practice medicine in Alabama, even though "black folks needed doctors." Dorine's first violent reaction—"We got to fight them . . . crackers. . . . Blow their Goddamn brains out"—soon turns to resignation. Her brother, like Francie's mother in *Daddy* and Cora's father in *A Short Walk*, counsels patience, insisting that "things will change" and that he plans to stay in the South "to try to change" them. Before she heads back to Harlem, Dorine makes her usual rounds of her hometown, and concludes that "Montgomery had not changed. Nor was it about to" (237–39).

Back in New York, Dorine finds she can no longer deny the changes the Depression has brought. Harry Brisbane, her common-law husband, demeaned by unemployment and dependency on her, sinks into despair and loses his grasp on reality. Soon he is dead. Poverty has made even thieves more desperate, and the relatively harmless pranks of Dorine's gang are being threatened by gun-carrying hoodlums. At the end of the decade, Dorine herself is arrested and sent to prison.

The first two books focus on the twenties; the third on the Depression; the fourth on the post-war decade 1945–55, when Harlem was in its final decline and the momentum of the civil rights movement was building in the South. When Dorine comes out of prison in 1944, many of the old gang are gone, dead, or dying. Formerly elegant apartment buildings are sordid tenements; the once-crowded Garvey's corner is occupied by a "small, light-skinned" man speaking to an indifferent audience (278). An old friend explains that while the war has "messed up with things something terrible," it has also brought employment to blacks in "defense plants, factories, even . . . department stores" (283). But for Dorine, change means "something's always messing up things" (283). She even experiences the death of Roosevelt as a personal loss, imagining that he might "have been able to help Harry" or to bring her son safely back from the war (298–99). Though she now recognizes that public figures and historical events affect her private life, she never considers that she might have an effect on history.

On what she calls her "absolute last visit to the South" (318), Dorine secretly brings her brother to New York to protect him from white reprisals for his work organizing sharecroppers to vote. Soon her own son, called "Son," recently released from the service, joins them. For a while Dorine has a family and a legitimate business, "a bar in the middle of Harlem," which she and Son operate (345).

The conclusion recounts the events following the Supreme Court decision of 1954 mandating the desegregation of schools: "black folks grinning and handshaking on back and front pages of all the newspapers"; a man named Malcolm X "shouting and pointing his finger" on Garvey corner; and children "starched and pretty, heads all proud, just a-marching" in Montgomery (348). Ironically, the effect of these events on Dorine's private world is devastating. She has just "closed a deal" for a new bar, when she hears the news of Rosa Parks's historic refusal on the radio. After talking to her niece in Montgomery, who tells her that the "bus boycott is going to be the biggest thing that ever happened in these parts" (351), she begins to fear that she is about to lose the hard-earned peace and prosperity she enjoys with her son: "After all these hundreds of years, a fool woman sits in the front of a bus in Montgomery and it's about to mess the hell outa my life" (351).

Her fears are justified. Her brother, having returned to Montgomery, decides to stay, and her son prepares to join the fight. Then Sonny dies. Dorine is alone, mourning for what she has lost and puzzled by the pending change. When her son rushes out to her car to tell her that her brother has "thrown caution to the wind" and is "joining up with Reverend Martin Luther King," Dorine lashes out: "Who in the fuck is Martin Luther

King?" (359). With these words, echoing her naive response to the news of Lindbergh's flight almost thirty years before, she drives away, roaring up Seventh Avenue.

The socio-historical backdrop of Dorine Davis's life is dramatized by narrative voices that are both in the time and looking back at it. Francie's unmediated and unqualified voice in *Daddy* allows readers to connect her narrative with their own time directly and with ease; the dual perspective of the first-person and third-person narrators of *A Short Walk* permits readers to distinguish between Cora's perceptions, informed by personal need, and the more public version of events reported by the objective narrator. By having Dorine narrate the entire novel in first person, alternating between her immediate, spontaneous voice and a more informed voice that looks back and evaluates the past from a time that is clearly after the civil rights movement, Guy achieves a dual perspective that holds the novel's competing worldviews in unresolvable tension.

As the knowing Dorine relates the exploits of her younger self, she does so primarily in her own youthful voice, making clear that, until the beginning of the civil rights movement, she was largely ignorant of public events and determined to protect herself from their assaults. Although she does not interfere with that younger self's assessments, the voice of the elder Dorine reveals that she now knows better. For example, she alludes to days before the movement when the only legitimate work available to black women was domestic service: "In those days black women didn't work in factories, nor at lunch counters, except those few in black neighborhoods. We didn't work as nurses in hospitals, not even as attendants, nor as salesgirls in department stores. And we hadn't begun to dream of being clerks in office buildings" (52–53). In this same reflective voice, however, she calls attention to her own ambivalence about changes in public behavior and attitudes, explaining that "in those days white cops didn't need to hide their hate from us" (162); she marvels at a time when she and other black outlaws "didn't carry guns," since back then they experienced the world as "a wide-open place where we floated in our kind of freedom" (163).

Implicit in Dorine's historically distanced meditations is a recognition that there has been some progress—better jobs for black women—but that other changes, such as increased violence and racism going underground, may not have been for the better. In her naive narrative voice, which persists to the end of the novel, Dorine expresses no curiosity about the historical context of her life, and in fact she seems to enjoy flaunting her ignorance. For example, shortly after arriving in Harlem from Montgomery, hearing street speakers talk of freeing Marcus Garvey and continuing the back-to-Africa movement, Dorine innocently declares: "Africa? What damn Africa? Here I had just got to New York and loved it, and these folks talking about leaving? I looked around at the hot faces, the thrown-back heads, the mouths tight and determined. I had come to find West Indians, and instead found a bunch of Africans" (40).

Ignorant of the history of her own race, Dorine does not know enough in the 1920s to have been affected by Marcus Garvey or to be impressed when she meets W. E. B. Du Bois. But the way she shapes her story and the tone of her older voice indicate that public events do eventually come to have meaning for her. For example, after World War II, when her brother sees her apartment in New York, he notices that she has hung pictures of Du Bois and Booker T. Washington side by side. When he teases her about hanging these two old enemies "together to bring their points of view closer" (317), she promptly puts him down, pointing out that Du Bois is a friend, that they "used to hang out together in The Club," and that she "just had to have old Booker T., him coming from 'Bama" (317). Though she tells this story to emphasize the personal aspect of her relationships and loyalties, she also reveals in her older self the knowledge necessary to understand that her brother is making fun of her.

Naive Dorine in 1954 still does not value her brother's commitment to the incipient civil rights movement, and she still believes in staying out of public movements and controversy. But the moment of her brash dismissal of King may have been the last time that such innocence was possible, since in a very short time the whole country would know Martin Luther King, Jr., and many African Americans would enter the public arena and "start throwing caution to the wind" (324). At some time after the main action of the novel and after the civil rights movement, Dorine seems to have learned about the many competing factions within the black community, even in her own private life.

In her knowing voice Dorine introduces the chapter that relates the hardships of the Depression:

> Depression. A plague over us—the crumbling of our kingdom and more. More than the white gangs taking over policy in Harlem, more than the riots from Harlem to Detroit where black folks were getting their heads beat in begging to be able to work. More than the warped and twisted soul of the southland being ex-

posed to the world. More than the invasion of Ethiopia by Italy, and the riots taking place between Italians and black Americans on the streets of Harlem. It was that—and more. (247)

This is not the voice of Dorine of the 1930s who had bragged about being untouched by the Depression until it was about "to drive a crate" over her (188). Dorine now speaks from a time when she knows more about world events and their impact and about geography—Ethiopia, after all, is in "damn Africa." This more knowing voice, looking back with historical perspective at a time when external events could no longer be ignored, periodically breaks through, though it never dominates. The Depression, Dorine observes in her knowing voice, was "a hell of a time for a marriage to be played out. But that was the time ours had to be played out—to the sorry end" (248). She begins to admit to "knowledge, long known and long hidden" (249), both about the public world and her own private life—that the Depression "was hovering over us all" (251) and that the man she lives with is seriously ill. This historically conscious voice, informed by events outside the novel, ties Dorine's story to the time of the novel's publication, when racism was less overt, violence more prevalent, and some successful women more concerned with being slender than with taking care of those they left behind. Recalling how men admired her when she first arrived in Harlem, Dorine observes that "Being a little plump in those days didn't move women to tears" (18). Dorine's discriminating voice is that of one who understands something about her place in history.

The elder Dorine accepts her own naivete without affirming it and celebrates the times without validating the self-indulgence they spawned. The way Dorine relates the story of her private life against the backdrop of the people and events that created the civil rights movement calls attention to her inability to see that something momentous was about to happen and suggests that those who concentrate on private matters, whether in 1927 or 1983, are likely to be blind to larger public issues. In this way the novel prompts readers to take a measure of their own time and perhaps to see parallels between Dorine's experience and their own. For Dorine and many of her counterparts in Harlem, a better life began when they broke away from the impoverished, racist South. At a time when the world changers had decades to wait to see their goals realized, individuals like Dorine and her cronies could not even imagine what would be necessary to right the wrongs they deplored.

For middle-class black readers in the 1980s, a better life may have begun thirty years before with the Montgomery bus boycott. When this novel was published in 1983, not only were many of the goals of the civil rights movement unmet, but its gains were being threatened, and many who had profited from the movement had hardly more understanding than Dorine about how they might play a role in fulfilling its dreams. Once again there were questions about whether private successes or public actions had more effect on society and which voices from the past could best serve as guides in the present. Dorine's two voices—one speaking with the immediacy of lived experience, the other looking back with historical perspective—offer two ways of measuring such times. They do not blend or harmonize. The novel does not resolve the contradictory forces of the public and private, does not tell us what to do. But in carrying us back to earlier times, it raises vital questions for the last years of the twentieth century.

At the end of the novel, Dorine, in her knowing voice, grieves for what she has lost and glorifies the dreams that created the golden age of Harlem: "We had come to Harlem, Sonny and I. . . . We had taken a stand in a place we loved. We had lived, worked hard to add to its shine. We had . . . joined up with others who had shared a dream" (358). And in that same voice she celebrates the "world-changers" she once ignored (358), even calling for a memorial to Garvey: "Would someone ever think with pride of a smooth-faced West Indian—a millionaire—who raged from Washington to his island, demanding something he called dignity for his people?" (359). Immediately after this lyrical reminiscence, Dorine shifts to her brash, irreverent voice for the novel's last words, a paean to Harlem and "Seventh Avenue—once the grandest, damnedest avenue in all of New York City" (359).

By dividing her vision of the Harlem Renaissance, the racist South, and the nascent civil rights movement in Dorine's two voices—her naive, largely private voice of the past and her present knowing public voice—Guy maintains unresolved tension about the role the past plays in the present and the relation between the public and the private. Though the narrative is mediated through Dorine's voices, her way—perhaps one still unsung—is one of many that the novel poses as having contributed to the civil rights movement. In the end, other fictional characters, as well as historical figures, enter the narrative as Dorine expresses confusion about how her life relates to theirs. Dorine watches Malcolm X on a street corner criticizing King's strategies; she calls home to find out more about Rosa Parks; she listens with dread when she learns that her brother and son are "joining up with Reverend Martin Luther King" (359).

Sassy, defiant, and incapable of assuming the submissive, passive role that allowed many Southern blacks to survive, Dorine and her kind, in leaving the South, may have saved their own lives. When World War II is over, Dorine's son explains why he did not go back: "I keep telling you, if I had stayed in 'Bama, I'd have been dead—and more'n likely taken a mess of crackers along with me" (321). It is not hard to imagine Dorine doing the same. In Harlem, however, Dorine could earn the money to meet her family's basic needs and to educate her brother and her son, as well as nurture the aspirations, hope, and energy that would eventually feed the civil rights movement. Her actions were not, of course, intended to bring about social change; rather, "in the middest," motivated by personal ambition and driven by impersonal forces, she contributed her part willy-nilly to that change.

By the 1950s, the lure of Harlem is gone, and Dorine's son and brother, having fed on the vitality of its past, eventually return to the South to join members of a new generation that would find another kind of excitement, another kind of dream, to be realized in the streets, the churches, and the jails of Montgomery and other southern communities—and on the mall in Washington.

Though Dorine is privately openhanded with family and friends, she does not extend her generosity to society at large. While she objects to oppressive social conditions, she never thinks of ways to change them. Other characters also fail to see the interrelation of private and public endeavors. While the novel leads readers to see how those who drove the flashy cars and wore expensive clothes paid for with stolen money contributed to the world changing that would culminate in the Montgomery bus boycott at the end of the novel, Dorine's family does not seem to understand the part she has played by educating her brother and sending money home. By failing to acknowledge that his mother has made his activism possible, Son is like others of his generation who imagine that the movement was their own creation. The revolution that became the civil rights movement was, this novel suggests, a many-faceted phenomenon, extending over several generations, and fed by social forces as well as by individuals like Dorine, who perhaps could not act directly and survive.

The two passages that echo the title of this provocative novel suggest two kinds of change which no one can escape: the inevitable aging that all people experience with the passage of time and the changes that come with history. On a visit to Montgomery, Dorine is amazed how much the children have changed: "Kids are the measure of time—or I could convince myself that it had been only the day before when I had run from the old place vowing never to return" (185). Dorine, who gets old, gains weight, and loses her sense of adventure, frequently acknowledges the effects of the inexorable passage of time.

And from some point after the final scene of the novel, she reveals that she has come to understand that underlying the sensational public events she cannot ignore are more subtle factors determining her life. In an attempt to explain the behavior of a successful young man who has renounced his own mother, a friend observes to Dorine: "These kids is coming up at a different time, under different sets of pressures. No matter what we wants, kids ain't as much a measure of their folks as they the measure of their own time" (305). In the interim between 1955 and the time of the telling, clearly the post-movement days, Dorine has been educated by events and by her own son to recognize that measures of time are both personal and historical and that the two are interrelated.

There are many similarities between *A Measure of Time* and *The Color Purple*, published the year before. Dorine, like Celie, tells her own story, spends her childhood in the racist South, and endures being repeatedly raped by an older man. Both heroines eventually escape from tormentors and achieve financial independence; the abusive male characters are tamed by the women they once dominated. In the end, both women are well into middle age. More significant than the similarities of the two novels, however, are the differences. While Celie stays in the South, Dorine, in going to Harlem, steps into the public arena. Though both novels are concerned with change, Celie's progress and transformation take place entirely in the private sphere while Dorine, continually on a collision course with history, adjusts to changing conditions.

A Measure of Time received generally positive but relatively sparing reviews. In comparison to the attention given to Walker's best-selling novel, it has been practically ignored, and though it is still in print, it is rarely discussed by those concerned with black women's fiction. That *Measure* was largely ignored in 1983 may well be a measure of a time when many readers preferred the lulling effect of *The Color Purple* to Rosa Guy's more unsettling narrative that dramatizes the consequences of ignoring the public dimension of private life.

CHAPTER FOUR

PRIVATE LIVES BEFORE THE MOVEMENT

Alice Walker's *The Third Life of Grange Copeland* (1970) and Toni Morrison's *Sula* (1973) cover the years from about World War I into the 1960s or later. Spanning these crucial decades, they chronicle change in the lives of isolated individuals whose personal dramas are initially played out on the fringes of history but eventually collide with historical events including the civil rights movement. The linear form of *Grange* presumes positive historical progress, not unlike that of *Jubilee.* Though the narration of *Sula* is also linear, its chapters entitled with specific dates ranging from 1919 to 1965 moving forward through time, it is also circular, as characters return again and again to the past. In the end, some of its characters, the unnamed "young ones" (166), are caught up in the movement, but those fearful of change who limited themselves to private experience find that progress has destroyed both the compromises they have rigged and the old community of suffering. The novel does not lament the changes but laments that for some, doomed to circle back repeatedly to what they have lost, change has come too late.

Grange and *Sula* span virtually the same years: *Grange* opens in 1920 and ends around 1960, after the civil rights movement is well under way. Beginning in 1920, *Sula* follows the characters up to 1965, the culminating year of the movement. In *Grange,* the young generation is represented by heroic and attractive young civil rights activists. In *Sula,* the careless young, referred to only indirectly, mock their mothers and thoughtlessly abandon the community that nurtured them. Rather than getting an education, caring for the old, or working for civil rights, they seek instead its prizes, including the key to the cash register in the dime store. *Grange* exposes the devastating consequences of the racist practices of the twenties and thirties on the lives of rural Southern blacks and traces the beginnings of the struggle against such practices. *Sula* examines the toll that racism takes on the lives of members of a small Northern community. Walker extends Grange's story into the early years of the movement, highlighting in the end the growing social consciousness of Ruth Copeland, whose young life is on a collision course with history. Morrison's novel, on the other hand, concludes as Nel, now an older woman caught unaware by historical change, confronts her losses and the sadness of her diminishing life. Nel finds that public progress has great private costs, even for those not concerned with history or society.

Unlike the Harlem novels, *The Third Life of Grange Copeland* and *Sula* focus on people whose marginal lives are hardly affected by the Depression, which happened not so much to them as to people with bank accounts, mortgages, and jobs to lose. Although he travels to Harlem in the 1920s, Walker's protagonist, unlike the heroines of *A Short Walk* and *A Measure of Time*, fails to find a place for himself there and eventually returns to the South determined to create a sanctuary that whites cannot invade. Ironically, the whites who eventually do invade Grange's world are young civil rights workers advocating an integrated society. In Harlem, under the leadership of Du Bois, Garvey, and A. Philip Randolph, the struggle against racial discrimination had already begun in the twenties; in the rural South, and in isolated Northern communities like Medallion, Ohio, what was to become the civil rights movement was not felt until the fifties, and in some places the sixties. For many older people who lived through the movement years in such places, the changes had little practical effect.

While *Grange* concludes by highlighting the possibilities for members of the younger generation who will inherit the new world created by the movement, *Sula* focuses on characters born before World War I whose lives are mostly spent before the benefits of the movement are felt. In *Grange*, characters who live largely in the private sphere confront and are changed by those whose lives are defined by public action; in *Sula* the characters live mainly in isolation, unaware of how they might participate in or benefit from the larger world.

Grange Copeland, Alice Walker's first novel, was published in 1970, the same year as Morrison's *The Bluest Eye* and Louise Meriwether's *Daddy*

Was a Number Runner. Though Walker's first novel suggests that many of the issues of the movement are unresolved and its work far from complete, its forward-looking conclusion implies that progress will continue. As an activist herself and a product of the movement, Walker, only twenty-six years old when *Grange* was published, evokes the optimism of her own generation fired by recent victories, contrasting their sanguine attitude with the resignation and the self-destruction of those that came before. Morrison, more than forty when her second novel, *Sula,* was published three years later, focuses, as she did in *The Bluest Eye,* on the tragic lives of her parents' generation and grieves for their pain, goading readers to consider both the losses and the gains of the civil rights movement. Though the optimistic mood of some in the mid-sixties might have been tempered by 1970, there was still a confidence among the young that progress had been made and that, in spite of the tragedies and losses, there was still something to celebrate. By 1973, however, darker moods prevailed. The year 1973 was the morning after the jubilee, Nixon was seemingly entrenched in the White House, and the 1960s were coming under a more sober appraisal.

The Third Life of Grange Copeland

The concluding scenes of Alice Walker's *The Third Life of Grange Copeland* are set in the peak years of the civil rights movement. Every night young Ruth, who considers Martin Luther King, Jr., a hero, watches the news and sees "students marching, singing, praying" (232). Ruth identifies with the student activists and argues with her grandfather, Grange Copeland, that under King's leadership they can "change those crackers' hearts" (232); the old man, on the other hand, has no confidence that progress will be made in race relations. Baker County, Georgia, where Ruth and Grange live together on an isolated farm, has been invaded by civil rights workers—young black and white students agitating for integration and working to register voters. A straightforward, third-person narrative spanning some four decades, *Grange Copeland* relates a largely personal family history which at the end converges with the compelling events of the early civil rights movement.

The last half of the novel focuses on Grange's "third life" on an isolated farm literally fenced off from the white world, where he pursues private pleasures and raises his grandchild, whom he takes in after his son, Brownfield, kills her mother. The novel opens, however, in Grange's "first life" with his wife Margaret and their only child, Brownfield, in a run-down shack owned by the white man for whom he works "planting, chopping, poisoning, and picking in the cotton field" (7). Hopelessly in debt and degraded by poverty, Grange and Margaret succumb to endless rounds of hard work, carousing, violence, and repentence. After a week of exhausting labor, they spend Saturdays doing chores until night falls, and Grange heads for a nearby juke joint only to return hours later to beat and abuse Margaret. Contrite and subdued the next day, he goes to church and raises "his voice above all the others . . . in song and prayer" (13), but by nightfall, he and Margaret are fighting again. Monday morning they wake to start the cycle again. The only alteration in the pattern occurs when Margaret decides to join in the Saturday night escapades—drinking and, like her husband, finding solace in sex, first with her fellow workers and finally with her white boss. Grange's first life ends when he abandons his wife and child after she gives birth to a white man's baby; Margaret ends her own misery by poisoning herself and her infant, leaving seventeen-year-old Brownfield to fend for himself.

The story of Grange's first life occupies fewer than twenty pages at the beginning of the novel. The next ninety pages, more than a third of the novel, focus on Brownfield, who repeats the self-destructive patterns that shaped his father's first life. That Brownfield is trapped in the life that his own father eventually escapes is due in part to his ignorance of history and the ways of the world. While Grange knows how to read, Brownfield is illiterate. Having "never heard of the North Star," or of "which direction he should follow to go North" (31), Brownfield wanders for weeks and ends up at the Dew Drop Inn, where his father once drank and caroused with the proprietor, "fat Josie." For more than two years, Brownfield lives there, helping out with the business and alternately sleeping with Josie and her daughter Lurene, finding a semblance of manhood, as his father had, in sex. Then Brownfield meets and eventually marries Mem, Josie's pretty, gentle, and college-educated niece, who in their early days begins to teach him to read.

To escape his dependency on Josie, Brownfield blindly walks into the trap of sharecropping, and within three years he is hopelessly in debt. Yet he and Mem are happy for several years, until poverty and overwork drive him to drink and abuse his wife. Even though he sees "how his own life was becoming a repetition of his father's" (54), Brownfield continues to dull his pain with alcohol and spend his anger by beating his once-cheerful wife. Like his father, he suffers the oppression and cruelty of white bosses who treat him no better than a slave; like his father, he taunts his wife and seeks respite in Josie's consoling arms. Brownfield's life up to that point is both a re-enactment of his father's and representative of the many black sharecroppers who live much like their slave ancestors. As late as the 1940s, Brownfield is "traded" by one white farmer to another, "as if he and his family were a string of workhorses" (79). But unlike his father, Brownfield plays out his tragedy to a bitter end. While Grange had run away to avoid killing his wife, Brownfield eventually murders Mem, shooting her in cold blood as she walks toward the shack on Christmas Eve carrying fruit and gifts her white employer has provided for the children.

In contrast to the precise dating of *Sula*, the temporal context of *Grange* is suggested by a sparse and inconsistent use of dates. A "new 1920 Buick" identifies the time of the opening scene. Brownfield is ten years old. Some six years later, Grange leaves his wife and son in "the spring of 1926" (144). Four weeks after that, Margaret poisons her baby and ends her life; the next day, Brownfield leaves and, "after weeks of indecisive wandering" (31), ends up at Josie's place, where he stays "over two years" before he meets Mem in 1928 or 1929 (43). Since there is no reference to intervening events or the passage of time, readers must assume they marry around 1929. After "eight years" of marriage, Mem gives birth to Ruth, the last of her three daughters (72), but events occurring when Ruth is still an infant are specifically dated 1944 (88). Clearly, Walker intended Ruth's birth to be in 1943 or early 1944, the same as her own. The final scenes of the novel, when Ruth is sixteen years old, would have to occur in 1959 or 1960, dates that are consistent with events in the outside world, such as the voter registration campaigns of those years. The narrative, then, fails to account for some six years from the end of the 1920s to the mid-1930s. The twenties fade imperceptibly into the Depression, and World War II comes and goes, but except for Grange, who apparently learns about public events from street speakers in Harlem, none of the characters seem aware of what is happening beyond their own narrow world. The chronology of Grange's life is also slippery, and except for an episode in which Grange accidentally murders a white woman, none of his experiences during what was well over a decade in Harlem, his "second life," is dramatized.

When Grange returns from Harlem, he comes to a static, ahistorical world. After marrying Josie and, with her money and what he has saved from hustling and stealing, buying a productive, secluded farm, Grange determines to create a world that is independent "from whites, complete and unrestricted," where he can live in "obscurity from those parts of the world he chose" (141). Determined to be self-sufficient, he produces his own food and wine and teaches his granddaughter what she will need to survive "*whole*" in a hostile world (214).

In all but the final scenes of the novel, characters show no awareness that they might be affected by or affect history. Not one fights in World War I, participates in the Garvey movement, or loses property during the Depression. Except for a minor character mentioned in passing (190), none of the men is involved in World War II. When Grange heads for Harlem in 1926, Marcus Garvey is already serving time in the Atlanta penitentiary. For the first two-thirds of the novel, the characters have no public existence: no birth certificates, no draft cards, no bank accounts. In Grange's first life, neither he nor the other characters intersect with history; during his second life in Harlem, Grange observes but never participates in public events; in his third life, when history in the form of the civil rights movement literally comes knocking at his door, he attempts to keep it out.

Yet in spite of his avowed commitment to protect Ruth from the outside world, Grange inexplicably instructs her in the lessons of history—presumably learned during his years in Harlem—and works to provide her with the resources she will need to go to college. The omission of any dramatization of Grange's encounters with activists and street speakers in Harlem leaves readers to wonder how "a small dog from the backwoods," a small-time street hustler, selling "bootleg whiskey, drugs, and stolen goods" managed to learn the lessons that he later teaches Ruth (144). He lectures her about "big bombs, the forced slavery of her ancestors, the rapid demise of the red man" (138). He provides her with books and builds a substantial bank account in her name. And he eventually relates to her all that he learned during his years in the North: "There were days of detailed description of black history. Grange recited from memory speeches he'd heard, newscasts, lectures from street corners when he was in the North, everything he had ever heard" (138).

When Grange first begins to educate Ruth, he cannot conceive of a world in which racial harmony is possible, for the first lesson Grange "learned" in Harlem and attempts to teach Ruth is to hate white people. He imagines that his decision to rob and then kill a pregnant white woman by allowing her to drown while cursing him with her "last disgusted breath" is the act that turns him from self-destruction to self-preservation (152). Grange's conviction that to free his manhood he had to kill "whatever suppressed it," and that the murder gave him a "passionate desire to live," lasts until the end of his life (153). He continues to associate violence with change, and when Brownfield murders Mem, Grange begins his third life by taking responsibility for his granddaughter. Still another murder will free her from this violent past.

In their long talks, Grange speaks of the liberating power of violence. Because he never tells her about the violence he committed in New York, Ruth associates murder only with her mother's death, and she is never persuaded that violence could ever produce a happy outcome. Grange, on the other hand, believes that the murder of the white woman has allowed him to direct his hatred "in the right direction" (155), as well as provided him with the money that allowed him to achieve economic independence from whites: "At last, he was free" (156), the narrator concludes. By inverting King's famous "free at last" refrain from the "I Have a Dream" address delivered at the March on Washington in August 1963, Walker calls attention to the distinction between Grange's and King's positions, and she keeps the issue in the forefront of the narrative by having Ruth continue to challenge her grandfather's militant and separatist position. No matter how vehemently Grange puts forth his view that all whites are evil and to be avoided and destroyed if necessary, Ruth continues to propose alternative views. When Grange argues that "the black man must be friends to every other of the downtrod, especially if he's a man of color," Ruth argues, as King did toward the end of his life, that poor whites are just as "downtrod" and refuses to "see what their white has to do with it" (175).

Though she very early comprehends the destructive power of racism, Ruth learns that racism is not limited to whites. As a sixth grader, she receives a new world history textbook—new to her but previously used and discarded by students in the all-white school. Inside the cover of the book she finds a drawing of "The Tree of the Family of Man" (185), which features white people at the top and the label "Scientist" (185); at the bottom, after Orientals and American Indians, is the caricature of a black man as a savage, with a bone through his nose, sitting in front of a pot of boiling water as if he were waiting to cook "a visiting missionary" (186). Written in the hand of the white student who had the book before are the words "a nigger" (186). At that moment her teacher, a black woman, accuses her of not paying attention to what are clearly the empty platitudes that pass for her history lesson: "You're just like the rest of them. . . . You'll never be anybody because you don't pay attention to anything worthwhile!" (187). By "them," the teacher clearly means uneducated, lower-class blacks.

As the years pass, Grange and Ruth continue to argue over whether any whites can be trusted or whether racists will ever change. Ruth is an adolescent when public history invades the narrow world she shares with Grange, creating the possibility that she will find a "way out of no way."[1] Her disagreement with her grandfather intensifies when young civil rights activists arrive at their door to register Grange to vote, and as they watch the reports of the civil rights movement on television, Grange focuses on "the ugly cracker faces," while Ruth sees "nothing wrong with trying to change crackers" (232–33). And Grange foresees danger and disappointment for the young activists and for King himself: "He felt about them as he felt about Dr. King. . . . He wanted to protect them, from themselves and from their dreams, as much as from the crackers. He would not let anybody hurt them, but at the same time he didn't believe in what they were doing. Not because it wasn't worthy and noble and inspiring and good, but because it was impossible" (241).

Like Randall Ware in *Jubilee*, who uses militant language while setting an agenda similar to that of the nonviolent movement, Grange takes somewhat contradictory positions. For example, he admires King, but talks like Malcolm X; while he insists that he sees no hope that the movement will create change, he knows that change is in the air. By continuing to hold out hope that Ruth has a "chance," he encourages her to see what he cannot. Readers, however, who know how the events of the movement played out, may recognize that Ruth's chance will come when her private life merges with the public events swirling around her. Like Mr. Coffin in *Daddy*, Grange, in spite of his pessimism about the future, prepares Ruth to be ready in the event that times change: "Maybe something'll turn up. Things change. . . . there have always been black folks fighting for better. Maybe their ranks will swell till they include everybody" (195).

Like Dorine in *A Measure of Time* and Cora in *A Short Walk*, Grange has acquired money and the freedom that comes from economic independence. But something more than money will be necessary for Ruth to make that final leap into history. Since being released from prison, the crazed Brownfield is obsessed with the idea of taking his daughter away from Grange, "not because he wanted her, but because he didn't want Grange to have her" (227). In the end, Ruth's salvation depends on a final and definitive act of violence. When a corrupt judge awards custody of Ruth to Brownfield, Grange murders his son and insures his own death. Fleeing the scene of the murder with police close behind, Ruth observes that they don't have a chance. Grange's response—"I ain't . . . but you do" (246)—is his final contribution to her future.

By the time *Grange* was published in 1970, both King and Malcolm X, the leaders of the opposing factions of the movement, had been assassinated; the jury was still out, the issues were still being debated, and there was little consensus about what the next step in the struggle for social justice would be. Many of the young activists of the early movement days had become disillusioned and dropped out of the public arena. The late 1950s and early 1960s television images of young people bravely facing their antagonists had been replaced by scenes of rioters destroying whole sections of American cities. The messages of love and the dreams of King and his followers had been replaced by the militant appeals of the black power advocates whose separatist visions were more compatible with Grange's views than with the idealism of the young people Ruth admires. But the novel does not definitively take sides. *The Third Life of Grange Copeland* ends with unresolved questions about just what constitutes effective social change and about what could be done to improve the lot of the disadvantaged.

When Walker was completing *Grange* in the late 1960s, she may have had the same hopes for Ruth that she had for herself, and she perhaps did not know what was possible. The focus of the novel's ending, then, is not on the specifics of Ruth's future but on this old man, who, having transformed his own life and proved that it is possible for people to change, has used that life to give Ruth a chance. He imagines that his "mission" is to prepare her "for some great and herculean task, some magnificent and deadly struggle, some harsh and foreboding reality" (198), even though he cannot conceive of what that struggle will be. Ruth, on the other hand, knows she wants to go to college and to be part of the movement.

By leaving Ruth's future in the realm of possibility, however, Walker allows readers to imagine a future for her: to spin out a happy-ever-after ending for this girl who goes to college, joins the movement, and perhaps creates a life for herself in the public arena. Since Grange's last words to her—that he does not have a chance but she does—have some authority in the novel, one is tempted to read Ruth's story in the context of the time in which it is set and to spin out such a positive scenario. But readers in 1970, when the country was still reeling from riots, campus violence, assassinations, and Vietnam—and still bracing for more—knew that whatever happens to Ruth, the issues she debates with Grange will remain alive and unresolved in the culture at large. The ten years that intervene between the setting of the end of the novel and its publication may have seen the end to legal discrimination, but little had been done to alleviate prejudice and racism or to resolve the debate between the militant separatists and the pacifist integrationists.

The final passage of *Grange*, however, like the ending of *Sula*, focuses on the pain of one who will never benefit from the fulfillment of the dreams of the young. Suffering from a fatal gunshot wound and "rocking himself in his own arms," Grange murmurs these last words, "Oh, you poor thing, you poor thing" (247).

Sula

Begun in 1969 when *The Bluest Eye* was still in proof, *Sula* is Morrison's only novel with an introductory passage that she calls "a welcoming lobby" marking the line between "public and private, them and us." In an address presented at the University of Michigan in October 1988, Morrison acknowledged that she is embarrassed today for having accommodated her narrative to the white mainstream by beckoning her readers into the black culture, rather than snatching them up and throwing them directly into the confrontational content of the narrative. Nevertheless she points out that, in the context of the "extraordinary political activity" when she was composing the novel, she felt the need to move slowly. Analyzing this opening passage almost twenty years after it was composed, Morrison observes that it is "about the community" and that Shadrack serves to "cement" and Sula to "challenge" the community.[2]

Yet the word *community* itself does not appear in this opening passage. It begins "In that place" and continues, "once a neighborhood," which "wasn't a town," but was "their town" called "the Bottom" (3–6). The word does appear, however, in the novel's closing passage as Nel muses about the Bottom where she and Sula grew up: "Maybe it hadn't been a community, but it had been a place" (166). The Bottom that Nel remembers was a way that people were with each other that is no more. What Nel lacks, but what the novel provokes readers to provide, is the historical sensibility that would enable her to understand change. The crux of this novel may lie in the paradox of its subject—community—being the very concept that the novel calls into question. Equally provocative is the obvious way it creates temporal gaps. The scene in which Nel is wondering whether the Bottom was ever a community occurs in 1965, the year when there are only rumors about a golf course being built there; the opening scene, however, is some years later, four, perhaps more—after the golf course is in place. The social changes that took place during those years were retrogressive ones for African Americans, many of whom felt that the gains of the movement were diminishing as were expectations of continued progress.

This place that the novel beckons us into is already being dismantled in the opening paragraph. The time is the reader's present, and an omniscient narrator announces that the place being re-created is gone, its inhabitants displaced, its culture destroyed. Embedded in this late sixties nostalgia for a lost but not so distant culture is the other story that predates and is the precondition for all African-American experience: the story of slavery. Interestingly, Morrison does not mention in the Michigan address any embarrassment about what the narrator calls a "nigger joke" (4), also part of the prologue, that explains how the Bottom came into being when a white man tricked a black slave into accepting the inferior high land as payment for his labor by promising him freedom and a piece of bottom land. The slave, assuming that bottom land is the rich land in the valley, is surprised when the white man explains that the real bottom is the bottom of heaven, high up in the hills. By accepting land "where planting was backbreaking, where the soil slid down and washed away the seeds, and where the wind lingered all through the winter" (5), he set a precedent for capitulating to white exploitation that his descendants would continue for generations. The narrator tells the joke, however, ostensibly to make a point that could still be applied to the conditions of many blacks in America at the time that *Sula* was published: "Freedom was easy—the farmer had no objection to that. But he didn't want to give up any land" (5).

Readers of *Sula* in 1973 would have just lived through the years in which the dissatisfaction with economic repression and residual social racism made the gains of the movement seem increasingly problematic. By then it was clear that the freedom granted by the civil rights movement—the right to sit in the front of the bus, to use public facilities, and even to hold the keys to the cash register in the dime store—was relatively easy to bestow. But in those days, the forces of reaction had already set in to counter the possibility that those in control might have to "give up any land." Equality in seemingly trivial political matters was acceptable; economic parity was something else. And the twentieth-century equivalent of giving up the land was exactly what Martin Luther King, Jr., was calling for at the end of his life.

The characters of this relentlessly chronological and implicitly historical novel are largely unaware of history. Unlike *The Color Purple*, which covers some of the same years without reference to a single date, or *Grange*, which slides somewhat carelessly through the same years, *Sula* uses specific dates to announce each chapter: 1919, 1920, 1921, 1922, 1923, 1927, 1937, 1941, and 1965. After the lyrical prelude that evokes life in the Bottom at a time when a visitor might see "a dark woman in a flowered dress doing a bit of cakewalk" (4), the opening chapter, "1919," introduces Shadrack, who never recovers from the shell shock he sustains on a World War I battlefield. When he returns a crazed though harmless recluse, the community accepts his condition—the direct consequence of a historical event—but does nothing to alleviate his suffering. In subsequent chapters, characters occasionally encounter Shadrack, but more than two decades pass before the consequences of his trauma during World War I affect anyone else.

Shadrack has long controlled his fear and despair by creating National Suicide Day, which he celebrates alone since no one accepts his invitation to join him in a parade and to confront and dismiss the possibility of suicide. These annual marches, like the nonviolent demonstrations of the civil rights movement, eventually metamorphose into suicidal violence, rioting, and destruction. When Suicide Day of 1941 dawns unseasonably warm, some members of the community impulsively join Shadrack's parade. As they reach the edge of town, they come to the tunnel which, because of discrimination in employment, was built without African-American labor. In a spontaneous act of rebellion, they begin to destroy the "tunnel they were forbidden to build" (161); some die there when the earth caves in. The frenzy in the tunnel is, in a sense, a delayed consequence of official policies that denied the African-American World War I veterans and defense workers the full participation in American society they expected. Ironically the frustrations and rage of decades of discrimination expressed in Shadrack's followers' fatal assault on the tunnel erupt on the eve of what will be the first tentative steps toward the changes that would take place during the next twenty-five years.

Though he is present for and in one case the cause of the crucial events of the novel, Shadrack is not the central character or a member of the two families that are at the center. Most of the other characters in the novel are attached to one of two families. Presiding over a "household of throbbing disorder constantly awry with things, people, voices and the slamming of doors" (52) is the one-legged Eva Peace, whose brief marriage to Boy Boy resulted in three children: Plum, Hannah, and Pearl. Eva's household includes a light-skinned alcoholic man named Tar Baby and three neglected children she collectively refers to as "the Deweys." Also in and out of Eva's doors are her own gentlemen callers and her daughter Hannah's many casual lovers. Like Shadrack, Eva's son, Plum, is also a victim of World War I. Before the war, he had "floated in a constant swaddle of love and affection," only to return home hopelessly addicted to narcotics (45). But like Shadrack's, Plum's misery affects others, and when his mother Eva ends his suffering in 1921 by setting him on fire, she joins the casualty lists of the history that underlies the personal and private arena. Hannah's only child, Sula, grows up in the midst of this hubbub.

Helene Wright's "incredibly orderly" house, complete with lace curtains and a velvet sofa, is consistent with her bourgeois values. The wife of a ship's cook on one of the Great Lakes Lines, Helene is often at home alone with her only child, Nel. The heartbeat, if not the heart, of this novel is the story of the friendship of Sula Peace and Nel Wright that began in 1922, when they were "unshaped, formless" twelve-year-olds, and that was irreparably damaged fifteen years later, when Sula carelessly engages in sex with Nel's husband, Jude. The friendship, however, is still alive for Sula on her deathbed in 1941 and for Nel almost twenty-five years after that. The story of Nel and Sula is "political" in the way that it explores the most personal of human experiences—friendship, first love, shared guilt, and sexual intimacy—in the context of a community and of history. Their relationships with their mothers, with boyfriends, and finally with men are historically shaped by conditions that they do not understand, but which readers may supply from their own knowledge of the history of the time.

For Nel, a specific racial episode leads her to make decisions that affect the way she leads her life. As a ten-year-old child, she accompanies her mother to the bedside of her dying grandmother in New Orleans. As they board the Jim Crow train, the white conductor insults them even before they can find seats in the colored-only car. Accepting abuse, enduring the scorn of her fellow black passengers when Helene involuntarily smiles appeasingly at the insulting white conductor, and squatting in fields to relieve themselves, Helene and Nel survive the two-day trip. But by the time they return home, Nel has "resolved to be on guard—always" (22). Though she dreams for a while of faraway places, she never again leaves the town where she was born for fear of submitting herself to that other world where racism is more overt and cruel. To be always on guard against such experiences requires that she limit herself to a very narrow range of action and that she never take a stand against racism.

In the years that follow, Helene, herself part white, gives her daughter mixed messages about the acceptability of her racial identity. Though she claims to be glad that her daughter is darker that she is, Helene condemns Sula's mother for being "sooty" and urges Nel to pull on her wide nose to change its shape. Black, in Helene's mind, is decidedly not beautiful.

It is against her mother's wishes that Nel befriends Sula Peace. The only character in the novel to confront white racists overtly, Sula, like rioters destroying their own neighborhood, threatens them by hurting herself. Four white boys, sons of Irish immigrants, themselves ostracized from the established white society of Medallion, entertain themselves after school by "harassing black schoolchildren" (53), until one day Sula confronts them. As the boys approach, she puts down her books, gets out a knife, and deliberately slices off the tip of her finger, taunting them to imagine what she will do to them if they continue to bother her. Like her grandmother Eva, who, the novel suggests, allowed a train to run over her leg to collect the insurance money to support her starving children, she equates survival with self-mutilation. Sula continues to engage in self-destructive behavior by living an independent, defiant, and self-indulgent life, deliberately cutting herself off from those who care about her, separating herself from community.

The male characters are crippled by racist employment practices. Particularly poignant is the dilemma of young men who long for the man's work they are so capable of doing, while the world they live in will allow them nothing more strenuous than carrying trays and peeling vegetables. As he watches the white gang boss pick "thin-armed white boys" to build the road that he so much wants to build, Nel's husband, Jude, gradually realizes that black men will not be hired. His hope is soon replaced with "rage and a determination to take on a man's role anyhow" (82). And the only man's role available to Jude in 1927 is that of husband, father, and adulterous lover. Ten years later, resigned to the menial work that is barely adequate to support his family, Jude blames racism for his plight. When Nel greets him at the end of the day with an idle question about what he knows good, his response is pointed: "White man running it—nothing good" (102).

In 1937 the men of the Bottom are led to hope that they will be given real work building a tunnel under the river. At the end of 1940 they are still waiting, "with a strong sense of hope" for the work that will mean they will "not have to sweep Medallion to eat" (151). Living marginal lives, they do not make enough money to buy Christmas gifts for their children. And it is in this context that they finally join Shadrack on National Suicide Day.

Sula and Nel, Jude, and Sula's lover Ajax are members of the last generation to be legally victimized by Jim Crow. They are trapped by society's restrictions, their own limited visions, and even by the hope they use to keep themselves going. Hope, in this novel, is the ally of the status quo and of those who want to keep the inhabitants of the Bottom outside of the mainstream of society: "The same hope that kept them picking beans for other farmers; kept them from finally leaving as they talked of doing; kept them knee-deep in other people's dirt; kept them excited about other people's wars; kept them solicitous of white people's children; kept them convinced that some magic 'government' was going to lift them up, out and away from that dirt, those beans, those wars" (160). Like Alice Walker's Grange Copeland, who looks to "presidents" to change things, they wait for some external reprieve and never consider the possibility that they themselves might bring about change. They are pawns, not agents of history.

In spite of their common childhood and adolescence, Sula and Nel lead entirely different adult lives, but in both cases those lives are notable for their repetitiveness. In the ten years they are apart, 1927–37, Nel engages in the same round of domestic activities, caring for her husband and children and maintaining their modest household, while Sula moves from one sexual adventure to another, until the endless men "had merged into one large personality" that was always "the same" (120). The monotony of their claustrophobic lives is in part a function of their being outside of history.

For all the mystery surrounding Sula's death, one thing is sure: she dies alone, isolated from and spurned by the community, defiantly dismissive of the world beyond her own subjective experience, longing for something that belongs to the personal past. Her last words, uttered, the novel tells us, after death—"Wait'll I tell Nel"—belong to the world of girlhood, when the confidentiality of shared personal experience was the essence of life. Nel's final words, spoken after she has confronted her own exclusion from the new society brought about by the movement, revert to that friendship: "O Lord, Sula . . . girl, girl, girlgirlgirl" (174).

The final scene of the novel, set in 1965, begins with a description of change: "Things were so much better in 1965. Or so it seemed. You could go downtown and see colored people working in the dime store behind the counters, even handling money with cash-register keys around their necks. And a colored man taught mathematics at the junior high school" (163).

Those young and resourceful enough to take advantage of the new opportunities brought by the movement have gained freedom from some of the conditions that marred their parents' lives. Those of Nel's generation who survive are still living in the narrow limits of the pre-movement days. For them, the movement has come, in the final words of *The Bluest Eye*, "much, much, much too late." For Nel, "Hell ain't things lasting forever. Hell is change" (108).

The final episode relates how the community that was once the Bottom has "collapsed" (165). Whites are buying up the land, and there is even talk of building "a golf course or something" (166). The prologue, set after the "blackberry patches" have already been torn "from their roots to make room for the Medallion City Golf Course" (3), explains how the white people have driven out the blacks and bought up the land to build hilltop houses with a river view. Public funds have been allotted to tear down the very buildings that had once been the center of social activity: the Time and a Half Pool Hall, Irene's Palace of Cosmetology, and Reba's Grill. Much of this change, however, is not the result of social reform, but of economic forces. While whites have money to build luxury homes in the hills, build golf courses, and buy out the structures that once formed the center of a black neighborhood, young blacks, whose grandparents had outhouses and took in boarders to survive, can now earn enough money for televisions and a house of their own. So while African Americans have been granted the privilege of riding in the front of the bus, working in dime stores, and being put away in their old age in integrated nursing homes, they have lost their homes and the mutual caretaking that had insured their survival for so long.

The inhabitants of the Bottom, however, never seem to have had the spirit of a beloved community that made the civil rights movement possible in other places and may be necessary if its gains are to be sustained. Though they care for each others' children, bury the dead, and celebrate marriages together, they also accept the inevitability of violence in their lives. Sula and Nel carelessly cause the death of a small boy named Chicken Little; Eva murders her son, Plum; Sula watches, without intervening, as Hannah burns to death. Although the characters relate to each other personally, they never take history into consideration or work together consciously and deliberately to improve their lot. They never consider the possibility of combining public and private lives or of taking action in the public sphere to preserve and nurture the community on which they depend.

Members of the next generation rapidly abandon the one place they have ever known:

> The black people, for all their new look, seemed awfully anxious to get to the valley, or leave town, and abandon the hills to whoever was interested. It was sad, because the Bottom had been a real place. These young ones kept talking about the community, but they left the hills to the poor, the old, the stubborn—and the rich white folks. Maybe it hadn't been a community, but it had been a place. Now there weren't any places left, just separate houses with separate televisions and separate telephones and less and less dropping by. (166)

Until Nel's final confrontation with history at the end, which she quickly dismisses by retreating into her personal memories, neither woman ever indicates an awareness of the political or social forces that inform the context of their small lives. The structure of the novel sharpens its focus on the private consciousnesses of historically unaware individuals. By omitting the decade between 1927 and 1937 and the quarter century between 1941 and 1965, the novel conspicuously excludes most of what we think of as the Great Depression and all of the years that make up the modern civil rights movement. The Depression was, of course, a time when public forces profoundly affected the lives of most Americans, and during the years between 1941 and 1965 virtually all of the changes that have affected the lives of African Americans took place. The omission of these years from the novel reinforces its concern with the consequences for those who live without attention to and involvement in the public forces that shape their world. There were, of course, many black communities, and clearly the Bottom is one, with no chapters of the NAACP or other civil rights organizations. And most Southern black colleges—probably the one Sula attends—discouraged activism among the students. The novel is not about what Nel and Sula should have done, but rather about the consequences of their living as they did—intentionally or unintentionally—outside of history.

Morrison has observed that there is an unavoidable "conflict between public and private life . . . two modes of life that exist to exclude and annihilate each other." This conflict, however, is important, even desirable, since "the social machinery of this country at this time doesn't permit harmony in a life that has both aspects."[3] The time Morrison refers to here is the period following the civil rights movement, a time dominated by the policies of Richard Nixon and Ronald Reagan, with Jimmy Carter's well-intentioned stand for human rights in the international arena having little impact on the prevailing retrenchment from policies to insure racial justice here in America. In such a time, Morrison urges, socially responsible individuals must live with the conflict resulting from the irreconcilable demands of public and private life. Embracing the conflict in her own work, Morrison insists that her novels are "political" and have little to do with her "personal dreams."[4]

Unrelenting in her insistence that all of her novels are political, Morrison invites readers to seek the political dimension of seemingly personal stories. How then is *Sula*, a novel about people who never consider that politics has anything to do with them, political? To live entirely in the world of "personal dreams," as the characters in *Sula* do, is in Morrison's terms to "annihilate" the public, to render oneself ineffective in the larger world, and in the process to become vulnerable to those who control that world. To be excluded from politics is to be doomed to the limited world of the personal, which can sometimes be managed only through the public sphere.

In *Sula*, as in Morrison's other novels, history provides the medium in which private lives grow. Ignorant of history, the inhabitants of the Bottom are being used by it. They do not know what Morrison suggests is necessary in "this country at this time"—that not to be swept away by history, they must consciously play their part in it. Rosa Parks, who is exactly Nel Wright's age, did just that when she joined the local NAACP in Montgomery and began the battle that would end in her refusal to give up her seat on a public bus in 1955.

Readers of *Sula* in 1973 and beyond who bring the context of public history to these lives know more about those times than the characters do. They know, for example, that, while Nel imagines that her world in 1921 was "*full* of beautiful boys" (164), many from that lost generation of "boys," like Shad and Plum, had been emotionally if not physically maimed by their experience in World War I. And they know that others of that generation were doomed by the demeaning racist employment practices that would insure that in the eyes of many whites they would always be boys. Such readers also know that World War II brought different kinds of change, that by late June 1941—after A. Philip Randolph threatened a massive march on Washington—Roosevelt issued an executive order forbidding racial and religious discrimination in war industries and government training programs. Had she been aware of the world outside, Nel, who is at that time a high school graduate and only thirty-one years old, would have had opportunities to live a better life. But unlike Delta in *A Short Walk*, none of the characters of Nel's generation takes advantage of these changes occurring in the larger world, a world that is outside their community and consciousnesses.

By ending *Sula* in 1965, some years before the time of the beginning passage of the novel, Morrison invites readers to come full circle and begin again, to re-experience the narrative in terms of Nel's final question about whether the Bottom in particular and "the black people" in general had ever been a community at all. Readers who retrospectively span the scenes of this remarkably visual novel will see one character after another alone, lonely, alienated from others: Shadrack alone in his shack longing for the little girl with the purple belt who was his only visitor; Nel, much later, passing Shadrack on the road, each "thinking separate thoughts" (174); Plum enveloped in a drug-induced isolation; Eva torn from her roots and transplanted in a nursing home run by a white church; Ajax taking off for Dayton, where he will slink about airports alone longing to take to the air; Sula experiencing "a loneliness so profound the word itself had no meaning" (123) and later, at the moment of her solitary death, reaching out in imagination to tell Nel what it was like; and finally Nel herself walking along the road from Sula's grave, crying "a fine cry" with no bottom or top, "just circles and circles of sorrow" (174).

For all its lyrical evocation of the characters' personal dreams, *Sula* does not finally affirm the longings that feed them. Rather, it denounces the conditions and the limited perspectives that kept Sula, Nel, Shadrack, and Jude locked in the personal, subjective world of their own fruitless dreams.

In 1973, it would have been virtually impossible for readers of a novel like *Sula* to be oblivious to the public history of years that saw the passing of the Voting Rights Act; the escalation of the Vietnam War, with its disproportionate number of African-American casualties; Lyndon Johnson's announcement that he would not seek re-election; the assassination of Martin Luther King, Jr.; increasing outbreaks of urban riots; the election and re-election of Richard Nixon; and the systematic attack on the advocates of black power. With all this came the rapid retrenchment from public policies intended to combat racism. At the same time the African-American activist community was becoming increasingly fragmented. Within the novel, events of those years are only suggested by the brief description of the final destruction of one black community, as the bulldozers, paid for with public funds and therefore as agents of public policy, come in and level the last of a place known as the Bottom.

CHAPTER FIVE

FROM DESEGREGATION TO VOTING RIGHTS

There are at least four categories of activity that make up what we call the civil rights movement: the slow, long-term work of organizing communities, fighting through the courts, registering voters, and finally achieving political power; the more intense, concentrated work of mobilizing people for relatively short-term campaigns, such as those in Selma and Birmingham; the spontaneous acts carried out by isolated individuals or small groups, such as the sit-ins in Greensboro, North Carolina, that serve as catalysts for other events; and finally, the unpredictable, sometimes capricious acts of violence—committed by both sides of the struggle—that came in the wake of more planned and structured occurrences. In the periphery of the organizing and mobilizing activity were masses of people who had little idea about the substructure of the movement but who were ready to give or raise money; participate in a demonstration, a march, or even a riot; call on unregistered voters; or teach in a freedom school.

In Toni Morrison's *Song of Solomon* (1977), the conflicting agendas of the civil rights movement are embedded in and central to the text. In Kristin Hunter's *The Lakestown Rebellion* (1978), citizens of an all-black community organize and mobilize for a significant, but entirely local, protest, using the methods of the nonviolent movement; in Ntozake Shange's *Sassafrass, Cypress, and Indigo* (1982), one character, almost incidentally, helps raise funds for the movement; and in Shange's *Betsey Brown* (1985), a family is temporarily thrown into turmoil over a school busing order and the father's surprising decision to take his children to a civil rights demonstration.

The four novels considered in this chapter are set principally in the years of the movement in communities where civil rights activity never made national headlines. The primary action of *Song of Solomon* takes place in a town much like Flint, Michigan;[1] *The Lakestown Rebellion*, in the imaginary first northern stop on the Underground Railway; *Sassafrass, Cypress, and Indigo*, in Charleston, South Carolina; and *Betsey Brown*, in St. Louis, Missouri. While these novels include characters from across the social spectrum and portray with varying degrees of complexity the tensions generated by the conflicting elements of the African-American community, they feature protagonists who enjoy middle-class privileges and are practically exempt from the indignities of racism. All are narrated in the third person.

Song of Solomon is distinguished from the other three by its relentless but implicit critique of those who live apart from the concerns of the larger community and by its refusal to offer easy solutions. In *Lakestown* the characters' personal conflicts vanish as they shift their attention from private concerns to a common public cause. The chronological narrative of *Sassafrass*, interwoven with letters, folklore, recipes, and poetry, concludes with rather facile solutions to the characters' potentially serious difficulties; while in *Betsey Brown* conflicts generated by the characters' fundamentally different political positions are swept away as they follow the dictates of personal, largely sexual affinities.

In *Sula*, Morrison omits any reference to events occurring during the Depression and skips over the quarter century from 1941–65, the years of the modern civil rights movement. *Song of Solomon*, published four years later, in 1977, is set in the very years that are excluded from *Sula*, and, except for the first sixty pages and occasional accounts of past events related in dialogue, the action takes place precisely in the peak years of the civil rights movement. Its climactic scenes coincide with the most dramatic and devastating episodes of the movement's most memorable year, 1963. The action of *The Lakestown Rebellion* also has resonance in the historical world. Set in a community with a long history of social protest, it takes place in July and August 1965, just before rioting in Watts signaled the end of the nonviolent era.

Temporal indicators in *Sassafrass, Cypress, and Indigo*, though quite vague and somewhat contradictory, are sufficient to locate the action mainly in the sixties. *Betsey Brown* is set in 1959, a relative lull between two phases of the movement's crisis years.[2] But for people on the periphery of the movement, like the artists and performers of *Sassafrass* or the privileged middle-class characters in *Betsey Brown*, being *for* civil rights and occasionally raising money or attending a demonstration was all that was necessary to feel part of the action.

In *Song of Solomon* the movement emerges from the subtext into the primary narrative, as some characters see events in the outside world as having a direct effect on their lives, others as being irrelevant; in either case, their response is representative of attitudes that are informing the public dialogue and creating what one character calls "the condition our condition is in" (222). In *Lakestown*, the one character who has been mistakenly hailed a hero in the movement actually does find a meaningful public role to play in the local Lakestown battle when racism and exploitation are personalized by threats to his own home. Though the Lakestownians do not speak of movement leaders or the passage of the Voting Rights Act, which Lyndon Johnson signed into law during the rebellion, the novel affirms the methods of the movement by having the rebels sing movement songs and respect the principles of nonviolence.

Most of the characters in *Sassafrass*, absorbed in their private lives, are largely removed from the public fray. The one character who joins a dance troupe that happens to be raising money for the movement seems to be motivated mainly by a desire to please her boyfriend, who is "from a long line of civil rights activist doctors and preachers" (186–87), but in the end he urges her to leave the troupe and marry him. Though the protagonist's father in *Betsey Brown* believes in the goals of the movement, there is nothing in the novel to suggest that his civil rights activity is grounded in organizations or that he expects to do more than attend—albeit with his children in tow—occasional demonstrations.

Song of Solomon

Toni Morrison's *Song of Solomon* may seem a largely apolitical, ahistorical fiction that yokes African folktales, biblical allusions, logical causes, and magical forces to tell the story of the private lives of the ill-fated descendants of Macon Dead I, an ex-slave, and his wife, Sing, a free woman of mixed blood. This extraordinary story of the "Dead" family, filled with magic, myth, and improbabilities, is also tightly tied to realistic portrayal of ordinary events of everyday life: lovemaking, marital conflicts, food preparation, dinner table talk, Sunday afternoon drives, and ladies' teas. Readers caught up in the miraculous elements of the story or in the compelling details of the domestic drama may fail to note the historical dimension of the novel.

There is, in fact, a significant relationship between the private lives and the real events taking place outside, though it is all but hidden by the fictional text of the novel. At least since Lukacs identified Balzac as the greatest of the heirs of Sir Walter Scott,[3] literary historians have defined the historical novel not in terms of its temporal remoteness from the present but in its relating of the fictional lives of the characters to the social and political world of their time.[4] *Song of Solomon*, then, is a historical novel, but one of a new kind, a palimpsest in which the partial erasure of the fictional text reveals the historical one.

Throughout the novel, Morrison explicitly and implicitly suggests the underlying history over which these lives are lived. She does this in several ways: by referring to real people and to documented historical events; by dating the episodes carefully, indicating the birth dates of almost all the major characters and therefore how old they are at particular points in the fiction; by referring to dates that have particular significance in black history and to the pervasive racism of the society; and by suggesting in subtle ways some surprising parallels between the lives of the fictional characters and their historical "counterparts," a term whose implications will be clarified later. The historical text of the novel stands, indeed, as a reminder of the consequences of living as the characters do, with little awareness of the historical forces affecting their lives.

In a short essay, published in 1984 under the title "Rootedness: The Ancestor as Foundation," Morrison explicitly confirms this, insisting that it is her intention to use fiction "to point out the dangers, to show that nice things don't always happen to the totally self-reliant if there is no conscious historical connection," and she makes it clear that she values "art" that is informed by political vision: "I am not interested in indulging myself in some private, closed exercise of my imagination that fulfills only the obligation of my personal dreams—which is to say yes, the work must be political. It must have that as its thrust. That's a pejorative term in critical circles now: if a work of art has any political influence in it, somehow it's tainted. My feeling is just the opposite: if it has none, it is tainted."[5]

The political vision of *Song of Solomon* emerges from the tension between two characters' conflicting and equally ineffective responses to the world in which they live. The protagonist, Macon Dead III, called Milkman, is self-absorbed, unaware of the history of his family, his race, and his society. Until his thirty-third year, he lives a complacent and irresponsible middle-class life entirely in the private sphere. He spends his days working for his prosperous father and his evenings partying with his middle-class friends; making love to Hagar, his simple-minded cousin; or hanging out with his friend Guitar Bains. An urban, working-class man, Guitar is cut off from his roots and scarred by years of poverty and mistreatment by middle-class people, black and white. Though he thinks he knows "every public thing going on in the city" (46), he is limited by his ignorance of history and his failure to respond productively to the social injustice he deplores. Like Sethe at the end of *Beloved*, Guitar seeks revenge by striking out at innocent whites, but while her act is spontaneous and short-lived, Guitar's is deliberate and planned. He joins a group of vigilantes, the Seven Days, who respond to violent acts by whites against blacks by randomly choosing an anonymous white person to murder. Believing himself to be acting for others, he is in his ignorance consumed by hatred and vengeance.

The last half of the novel chronicles Milkman's struggle to discover his personal history (an effort that begins as he seeks the history of his own family) and Guitar's increasing obsession with revenge and violence. The lives of Milkman and Guitar reflect the limited lives of so many black Americans who for so long did not or could not find their way into the stream of history. The situations of the fictional characters at the end of the novel are emblematic of the ambiguous future of the black community in 1963 and beyond. Some, like Guitar, were torn between violence and reconciliation; others, like Milkman, were poised on the precipice of possibility. *Song of Solomon* is a novel about the dangers of trying to live entirely in the private sphere.

Much of the significance of the novel depends on the reading of the historical text under the private fiction. The poverty of the characters' lives is at least in part due to their ignorance of history. To suggest that those who remain unaware of the historical forces of their time—as the characters do—are doomed to live unsatisfactory lives, the novel subtly makes history shine through into the fiction, alerting the reader who is not imprisoned in a narrow private life to its presence, quietly calling attention to the great public events and the people who brought them about, just below the shallow world of the characters' private lives. Names of real people scattered throughout the narrative text, for example, help suggest the historical text beneath. The narrator refers to Charles Lindbergh, Father Divine, Franklin Roosevelt, and Harry Truman; various characters speak of Herbert Hoover, Albert Schweitzer, Eleanor Roosevelt, Elijah Muhammad, Sam Sheppard, Louise Beaver, Butterfly McQueen, Orval Faubus, and Adolf Hitler. Guitar warns Milkman not to "let them Kennedys fool" him (225).

A detailed chronological scaffolding both marks the events of individual lives and invests those events with a significance far beyond the personal. With more specificity than in *Sula*, the historical underpinning of the lives of the characters is suggested by explicit dating. The first scene of the novel—with the insurance agent Robert Smith perched on the top of Mercy Hospital, ready to fly, or die—occurs on February 18, 1931, Morrison's own birthday and the day before the birth of Macon Dead III, who gets his nickname "Milkman" around 1936, when Porter spreads the news that he has seen Ruth nursing her five-year-old son. The novel ends in October 1963 with Milkman in his thirty-third year, also on a precipice, ready to fly or die. Morrison dates the events both directly and indirectly: in a suicide note (Mr. Smith leaves a note announcing that he would take off from Mercy and fly away on his own wings "at 3:00 p.m. on Wednesday the 18th of February, 1931" [3]); in direct narrative statement (Dr. Foster "moved there in 1896" [4]); in dialogue (Macon refers to the time his father was given his name "in 1869" [53]); by reference to the ages of characters ("Milkman was twenty-two" when he hit his father [64]); by allusion to historical events that can readily be dated (the bombing of a Birmingham church that killed four young black girls [173]); and by reference to more obscure events in the historical past (Pilate tells Guitar that her father died the same year "they shot them Irish people down in the streets" [42]).

Specific dates mark the private events in the lives of most of the characters, including those whose stories are outside the main action. For example, the first Macon Dead, a young man when he left Virginia for Pennsylvania in 1869, was a successful farmer in the 1890s; he was murdered by white men around 1907. Dr. Foster came to Mains Avenue in the city in 1896, where he lived until his death in 1921. His daughter Ruth was born in 1901, married Macon Dead II in 1917, and had her last child, Macon Dead III, in 1931. First Corinthians, her eldest daughter, was born in 1919.

The second Macon Dead was born around 1891, and his sister Pilate was born four years later. A few days after their father's death, Macon and Pilate, on the run, fight about whether to take a treasure of gold they find in a cave. When Macon returns to the cave and finds that the gold is gone, he concludes that his sister has stolen it and fled. Macon was "already pressing forward in his drive for wealth" at the age of seventeen—around 1908 (28). By the time he reached twenty-five—around 1916—he was the owner of two rental shacks; and, as "a colored man of property" (23), he dared to court Ruth Foster, whom he married in 1917 (70).

Pilate was born around 1895. From the day she was separated from her brother, she traveled from place to place and supported first herself, and then her daughter and granddaughter. Like her brother, Pilate took advantage of the historical moment—Prohibition and the Depression—and profited from the troubles of others: she made wine and whiskey. Since the "crash of 1929 produced so many buyers of cheap home brew," Pilate made "a lot of money" (151), but she had no interest in accumulating wealth. What money she did save in that year, she used in 1930 to travel to the city where she found her brother and to set herself up there as a "small-time bootlegger" (150).

The events in the lives of Guitar and Milkman often coincide. At the age of "five or six" Guitar is in the crowd in front of the hospital the day before Milkman is born. They meet in 1943, and shortly afterward, Guitar takes Milkman to meet Pilate. On the same day Milkman has the first of three serious conversations with his father and falls in love with his cousin Hagar, Pilate's granddaughter. He first sleeps with her in 1948. In 1955 he strikes his father and precipitates their second serious conversation; on the same day, he discovers the origin of his name, tells Guitar about it, and listens to talk about the death of Emmett Till, the fourteen-year-old boy who was murdered in Mississippi after allegedly making a fresh remark to a young white woman.[6] On Christmas Eve 1962, Milkman breaks off with Hagar and has a major argument with Guitar. In late August 1963, Milkman has a showdown with the now murderous Hagar, and Guitar condemns him for mistreating her. In the next few days Guitar determines to avenge the murders of four black girls in a church in Birmingham. Milkman's final encounter with his father leads to an attempted robbery, Milkman's journey south, and the climactic final scenes of the narrative.

The dates of the fictional lives are inscribed over historical ones, which do not always show through into the words of the text. The first three dates mentioned in the novel—1931, 1896, 1918—have particular significance to the history of black Americans. The year 1931 appears in the opening lines of the novel. That year nine African-American youths boarded the Chattanooga-to-Memphis freight train only to find themselves accused of rape and their lives in jeopardy. The Scottsboro case, which is featured in the lives of the characters in *Daddy Was a Number Runner* and *A Measure of Time*, became a cause célèbre of the 1930s, keeping the issue of racial injustice before the public for years.[7] It was also in 1931 that the Detroit Temple Number One was established as a place of worship for black Muslims.[8] Though these outside events are not specifically mentioned in the text, as we shall see, they inform the context and have consequences in the novel.

In the second paragraph Morrison flashes back to two other dates—1896, when Milkman's maternal grandfather moved to his home on Mains Avenue called "Doctor Street," and 1918, the year when a few of the many black men who were drafted "gave their address at the recruitment office as Doctor Street" (4), the first instance of a specific reference to the impingement of history on the private fiction. Both years were marked by events that were catastrophic in the lives of black Americans. In 1896 the U.S. Supreme Court decision in the case *Plessy v. Ferguson* began the reversal of the progress that black Americans had made since the Civil War. The ruling condoned the practice of legislating separate-but-equal facilities for blacks and whites and initiated a long series of Jim Crow laws that were considered constitutional until the *Brown v. Board of Education* decision of 1954 and that were not changed in any practical sense until the Civil Rights Acts of 1964 and 1965.

In 1918 many of the 300,000 African Americans in the armed forces expected that, by fighting for "the democratic way of life," they would be granted full participation in that life. The thousands of black Americans who traveled north in search of well-paying jobs in war industries, like their brothers in uniform, soon discovered that their participation in the mainstream of American life was short-lived. When the war was over, many were fired and others were excluded from unions.[9] In 1919, race riots broke out in towns and cities all over the country.[10]

The first reference in *Song of Solomon* to the drafting of blacks in 1918, buried in a rambling explanation about the naming of "Not Doctor Street," may seem gratuitous. But in fact the lives of several of the characters were significantly affected by World War I. Macon Dead II, in his mid-twenties at the time, profited from the war by acquiring rental property at a time of growing need for housing. Tommy tells Guitar of the difficulty of killing with bayonets in the bloody battle of Belleau Wood—June 1918 (101). Another character, Empire State, went to France and entered a disastrous marriage with a white girl there. When Guitar tells Milkman about the Seven Days, he explains that it was started in response to atrocities suffered by black veterans in 1920—the castration of a private from Georgia and the blinding of a veteran when he returned home from France (155).[11]

The memories of the First World War and its aftermath are still present in the thoughts and lives of the characters almost forty years later. On the day in 1955 that Emmett Till was lynched, and Guitar's barbershop cronies (also members of the Seven Days) are imagining what will happen to Till's murderers, history comes to the surface: One man asks, "Remember them soldiers in 1918?" Another responds, "Ooooo. Don't bring all that up." But Guitar's older buddies—Porter, Nero, Walters, Freddie, Hospital, and Railroad Tommy—do bring it up, as they begin to "trade tales of atrocities" heard, witnessed, and experienced (82). When Guitar joins the Days and becomes one of the seven men committed to avenging the unpunished deaths of blacks, he is actually entering a stream of history that has its immediate roots in the racial conflict following the First World War and its distant roots in the acts of violence and the stream of discriminatory legislation that led to the race riots following the war.

Racism affects the everyday lives of characters confronted with discrimination in public facilities, transportation, employment, and housing. In 1931 Mercy Hospital—known to Southside residents as No Mercy Hospital—denies admittance to black patients and refuses to allow black physicians to practice there. Ruth Foster is admitted to Mercy where she gives birth to Milkman not because the policy of racial discrimination has changed, but because she goes into labor while watching Mr. Smith's leap from the cupola of the hospital building. In 1942, Pilate's daughter Reba goes into Sears to use the bathroom because there were only "two toilets downtown they let colored in," and "Sears was closer" (46). When Reba walks through the door and unexpectedly wins a diamond ring in a contest sponsored by Sears, the officials deny her the publicity that had been planned for the winner and instead publish the picture of the white man who won the second-prize. Years later, when Pilate tells the story of her adventures and difficulties living alone as a young girl, she relates how she first learned that black people in those days were not even allowed to ride some passenger trains and how, when she asked how they "get where they want to go," she was told that they "ain't supposed to go nowhere" (145). Also, Milkman's sister Corinthians, a graduate of Bryn Mawr, can only find work as a domestic servant. These undramatic examples of pervasive racism suggest the public realities and historical contingencies that substantiate, structure, and lend significance to the private fiction.

On a summer afternoon in 1936 Macon Dead II takes his family for a drive and explains his plans to build a community of beach houses for affluent black people. No one challenges the concept of a separate, segregated resort, or criticizes his intention to profit from it. The Dead family's acceptance of the separate-but-equal doctrine was consistent with that of many leaders of the civil rights organizations in the thirties, including

W. E. B. Du Bois and Mary Bethune.[12] Milkman absorbs his parents' seeming indifference to discrimination, and by working for his father and continuing to live at home in an affluent neighborhood, he avoids the direct experience of racial discrimination. And he does not respond with empathy to the many stories of racial injustice that other characters relate.

But if Milkman does not hear more about the evils of racism from his parents' conversations that summer Sunday afternoon, he encounters it repeatedly in the years to come. It is a long time, however, before he recognizes that he is as guilty of exploiting others as the white racists who haunt Guitar. On his first visit to Pilate's house—in the summer of 1943—Milkman learns about the murder of his grandfather. Later, on the same day, after having discovered Milkman's visit to Pilate, Macon seeks out his son and relates his version of what life was like back in Danville, Pennsylvania, before his father, Milkman's grandfather, was shot off the fence by white men who wanted his land. Although Macon does not answer Milkman's direct question about who killed his father (53), he does refer to the kind of racism that leads people to commit murder casually: "White people did love their dogs. Kill a nigger and comb their hair at the same time. But I've seen grown white men cry about their dogs" (52). When Milkman asks if his grandfather had been a slave, Macon calls attention to his ignorance: "What kind of foolish question is that? Course he was. Who hadn't been in 1869?" (53).

Milkman makes his first visit in 1944 to the barbershop where Guitar's buddy Railroad Tommy lectures him about all he will be denied simply because he is black: luxuries of all kinds, the freedom to buy a house wherever he likes, a high-ranking commission in the armed services.[13] Milkman, however, seems unmoved by the racial implications of the lecture and acts as if they have nothing to do with him. After they leave the barbershop, Guitar explains to Milkman why he cannot tolerate sweet desserts: his own father had been killed—his body cut in half—in a sawmill accident, and the boss's wife "came by and gave us kids" a sack of divinity candy (61). But Milkman remains indifferent.

In 1945, at the age of fourteen, Milkman is still concerned only with what touches him personally. When he discovers that one of his legs is slightly shorter than the other, he believes that he has a serious defect, perhaps polio, and the public figure that he identifies with is Franklin Roosevelt, but only because he imagines a kinship with the late president based on their shared handicap: "Even when everybody was raving about Truman because he had set up a Committee on Civil Rights, Milkman secretly preferred FDR and felt very very close to him" (62–63).

As a young man in his early twenties around the time of the first significant protests following *Brown v. Board of Education*, Milkman is still indifferent to problems of racial injustice. In a single day, he experiences a series of disturbing personal revelations that are juxtaposed in the text with a major public event—the murder of Emmett Till. Enraged at his father's violent treatment of Ruth, Milkman strikes his father for the first time. Macon defends himself by telling his son about what he imagines to be his wife's unnatural attachment to her father. Significantly shaken, Milkman leaves. Walking through the streets meditating on what he has been told, he suddenly notices that hordes of people are walking on the other side of the street in the opposite direction. Readers will soon know what Milkman does not know, that crowds are gathering to protest the murder of Emmett Till, an event that has the black community in an uproar. Though history has impinged on his world, affecting his community and the thoughts and actions of his friend Guitar, Milkman remains absorbed in his own personal world, in this scene literally walking against the tide of history.

As he walks, he suddenly recalls for the first time the event in his childhood that was responsible for his being called Milkman rather than Macon. He remembers having been in a small, green room; he was five years old, and his mother was nursing him. Someone saw them and laughed. The word got out, and he got a new name. This bothersome personal discovery now leads him to search out Guitar; he finds his friend with his cronies in Tommy's Barbershop, listening intently to the news about Till on the radio. When Milkman finally tells his friend what he has discovered, Guitar, uninterested in his relatively trivial personal problems, tries to turn the conversation back to Emmett Till, only to hear Milkman's egocentric retort: "I'm the one in trouble" (88).

Milkman *is* in trouble. Earlier in the day, immediately after he struck his father, Milkman is standing in front of a mirror studying his own reflection: "It was all very tentative, the way he looked, like a man . . . trying to make up his mind whether to go forward or turn back" (69–70). At this point the narrator observes that the decision that Milkman makes will "be extremely important," but the way he makes it will be "careless, haphazard, and uninformed" (70).

The narrative moves forward seven years, but Milkman steps back to dependency and irresponsibility. In the intervening years most of the episodes of what is considered the modern civil rights movement have occurred: the Montgomery bus boycott, the violent confrontations over the integration of public schools in Southern states, the founding of the Southern Christian Leadership Conference and the Student Nonviolent Coordinating Committee, the sit-ins, and the freedom rides—events that are never mentioned in the surface text and that are apparently absent from Milkman's consciousness.

On Christmas Eve 1962, Milkman is writing a letter to Hagar, breaking their fourteen-year-old relationship with a casual thank-you note. Still working for his father, he moves from one good time to another, from the lower-class Southside world that he visits with Guitar to the middle-class black neighborhood called Honoré. When Guitar accuses him of not being "a serious person" (104), of living for himself alone, not caring what is happening in Montgomery, of not knowing where he is going, Milkman insists that he is going "wherever the party is" (106). He recognizes that his life is "pointless, aimless" and that there is nothing he wants enough "to risk anything for, inconvenience himself for" (107). Milkman cannot even conceive of becoming involved in racial problems: "Politics . . . put him to sleep. . . . The racial problems that consumed Guitar were the most boring of all" (107). And this in December 1962.

Milkman spends his days in 1963 working for his father, passing time with his friend Guitar, and avoiding—often with Guitar's help—the vindictive Hagar who is determined to kill him. During the period from Christmas Eve 1962 to August 1963, Milkman is in limbo. He feels that he is not living as he should, but he does not know how to change, and he is apparently completely unaware of the sweeping events that have reached his own backyard. The only discussion about race that Milkman has occurs shortly after his final encounter with Hagar in early September 1963. Guitar is justifying the Days' practice of randomly murdering whites on the ground that all whites are capable of racial atrocities. Milkman counters by suggesting exceptions, not present-day allies in the 1960s' struggle for civil rights, but Eleanor Roosevelt and Albert Schweitzer.

The year 1963—the hundredth anniversary of the Emancipation Proclamation—may well have been the most dramatic and violent period of the civil rights movement. It was a time of boycotts, demonstrations, jailings, bombings, and assassinations; of relentless confrontations between Martin Luther King, Jr., and Bull Conner, and of regular communication between the White House and civil rights leaders. It was the year of the massive demonstration in Detroit, of the now famous "Letter from Birmingham Jail," the children's crusade, and the memorable March on Washington. It was also the time when the battle lines were drawn among various factions of the black community—and Malcolm X openly attacked King. In the months preceding the death of John Kennedy, events of the civil rights movement and its consequences dominated the news. Both fulfilling dreams and destroying them seemed equally possible.

In late summer and early fall of that year—the last days in the life of John Kennedy—the final scenes of the novel unfold far from the public arena. On August 28, more than 250,000 people gathered in Washington to participate in the most memorable demonstration of the movement. Hagar attempts to kill Milkman on August 30; shortly afterward, she loses her will to live and dies. On September 15, four black children are killed when a bomb explodes in their church in Birmingham, Alabama, an event that leads Guitar to go along with a plan to steal what Milkman thinks is a sack of gold because he needs the money to finance the vendetta for the bombing. But the sack turns out to contain human bones instead of gold. At this point the historical and fictional texts merge. On September 19, 1963, when Milkman and Guitar steal Pilate's treasure, they set in action a chain of events that lead Milkman to leave home and Guitar to follow him. Still convinced that he can find the gold, Milkman heads south to look for it. Until then, he is ignorant of his own history, of the history of his people, and of the momentous events swirling around him. Still ahead are the discovery of his personal history, the death of Pilate, and the final encounter with Guitar.

Within days after the robbery, Milkman sets out for Pennsylvania, where Macon and Pilate presumably found the elusive gold more than fifty years before, and later for the small town in Virginia, where he imagines Pilate may have left it. Following him is Guitar, diverted from his mission to kill a white person by his mistaken belief that Milkman has betrayed him. He sends Milkman the message that his "day" has come. The remaining episodes of the novel take place in late September and early October 1963, as Milkman pursues his own family history with the murderous Guitar following close behind.

Milkman's quest is successful: having found Circe, the old woman in Pennsylvania who tells him what she remembers of his grandparents and their deaths, he goes on to locate a remaining relative who helps him piece together the puzzle of his ancestry. This information, with the help of a song that the children still sing about his ancestors, leads Milkman to conclude—unlike those who tell him about the past—that his great-grandmother died of a broken heart and that his great-grandfather could literally fly. Again, Milkman does not consider what the attentive reader might, that the "Song of Solomon," itself a palimpsest, is actually a coded account of the escape of a runaway slave. When Milkman returns home to tell Pilate what he has learned, he expects her to feel as he does. What he does not know, however, is that in his absence Hagar, like her great-grandmother, has died of a broken heart. And Milkman has inadvertently re-enacted his grandfather's flight by seeking his own freedom at the expense of the woman he left behind.

What is left out of this provocative novel is any overt discussion of the political implications of Milkman's quest. In "Rootedness: The Ancestor as Foundation," Morrison insists that in a novel "what is left out is as important as what is there." It is the reader who must fill in the gaps, she tells us: "I have to provide the places and spaces so that the reader can participate. Because it is the affective and participatory relationship between the artist . . . and the audience that is of primary importance . . . to have the reader work *with* the author in the construction of the book—is what's important."[14]

Just as Milkman is questing for his own personal history—those acts of passion and violence in the past that have determined his personal present—so the careful reader of *Song of Solomon* is invited to seek the even more deeply hidden history, the implied historical reality that underlies and parallels the narrative of the personal lives of the characters. In the final chapter of the novel, Morrison notes: "How many dead lives and fading memories were buried in and beneath the names of the places in this country. *Under the recorded names were other names*" (329, emphasis mine). Obviously, the biblical names reverberate with meaning: Ruth, Corinthians, Magdalena, Pilate, Rebecca, Hagar. But there may be more to the naming than a simple correspondence between these characters and their biblical counterparts. Perhaps there are other names *under* the recorded names.

Consider "Guitar Bains." Under this reality may be another reality. Milkman accuses his friend of sounding "like that red-headed Negro named X" and suggests that he call himself "Guitar X" (160). There are indeed striking parallels between Guitar and Malcolm X, of which Milkman is unaware. Six years older than Milkman, Guitar would have been born around 1925, the year Malcolm Little was born. Both grew up in Michigan and had fathers who died violently—by having their bodies cut in half or almost in half—and both were separated from their mothers. Guitar was raised by his grandmother; Malcolm was placed in a foster home when his mother had a nervous breakdown. At about the time Malcolm X was released from prison and moved to Detroit, where he very soon became a leader in the Muslim movement, Guitar embraced vengeance as the appropriate response to racial injustice. Although Guitar does not understand Malcolm X's crusading spirit, he has taken on other characteristics: he has given up cigarettes and alcohol, rejected the idea that whites can work for the welfare of blacks, and adopted the antifeminism often associated with the Muslims (154–61). He has assumed the surface manner of Malcolm X, but not the substance of his mission.

Guitar does not want to change his name, however, and he argues that he does not care about what Malcolm X is trying to do. When Milkman reminds him that Malcolm changed his name in order "to let white people know you don't accept your slave name," Guitar retorts that he doesn't "give a shit what white people know or even think" (160). He has generalized his anger at particular white people to all whites. Guitar might be understood as the private, destructive counterpart to the powerful public force that was Malcolm X. In black folklore, the guitar is associated with evil.[15] "Under his name" may lie the suggestion of the man he might have been had he been motivated by creative rather than destructive impulses.

When Milkman and Guitar are planning to rob Pilate, Guitar tells him "Wanna fly, you got to give up the shit that weighs you down" (179). Guitar, however, is weighed down with vengeance, a burden at least as debilitating as the materialism and solipsism that hampers his friend. Having seen his own father cut in half in a sawmill and watched his own mother ingratiate herself before the white boss of his recently dead father, having been abandoned by that same mother and sent off to live in poverty with his loving but helpless grandmother, Guitar has known all too vividly a family history that has left him angry and determined to seek revenge. Instead of becoming involved in the larger civil rights movement that is everywhere around him, Guitar joins a secret society dedicated to seeking revenge.

If Guitar is the private, limited counterpart of Malcolm X, whose private counterpart, then, is Macon Dead III, called Milk by his friend Guitar? The name, of course, suggests infancy, dependency, weakness. But what other name lies under his? When Milkman goes south to the little town that his grandparents left in 1869, he goes hunting with a group of black men: one is named Luther, and the man who lends him a gun is named King. Under his name may lie another name. MILK. Remove the "I"—and with it the egotism and preoccupation with self that make meaningful life impossible—and you have MLK.

There are significant similarities bewteen Milkman and the apparently very different Martin Luther King, Jr. They were born at approximately the same historical moment, though King in 1929 and Milkman in 1931. Like King, Milkman had privileged, middle-class parents who expected him to take advantage of his social status to acquire property and financial security. But of course, the two men took very different routes through the same years. King graduated from college in 1948, the year Milkman begins the exploitive relationship with Hagar that eventually leads to her death; King took on the Montgomery boycott in 1955, the year that Milkman struck his father and demonstrated his total indifference to racial injustice by telling Guitar that he is much more concerned with personal injustices of his own life than in the senseless murder of Emmett Till. While King's busy agenda in 1963 included major protests in Birmingham, as well as his considerable contribution to the famous March on Washington in 1963, Milkman spent the same months indulging himself in parties and relishing the role of "one bad dude" capable of driving "a woman out of her mind" (301). Just as Milkman is preparing for his final showdown with Hagar, King reached what many would consider the zenith of his career with the remarkable March on Washington and the famous "I have a dream" refrain that was heard throughout the world. Milkman might be understood at this point as the moral and psychological obverse of Martin Luther King, Jr., a man as self-seeking as King was self-sacrificing, as oblivious to history as King was aware of it, as careless as King was caring.

If August 28, 1963, was a high point in the civil rights movement, a time when various factions sacrificed their special interests for the sake of one massive demonstration of harmony and common purpose, "buried in and beneath" this dearly bought coalition with a deliberately hidden factionalism and potentially fatal conflict. Morrison's narrative, then, is about the other side of the black community, the hundreds of thousands, perhaps millions, of black Americans who were absorbed in their own personal struggles and conflicts: the "Guitars" disabled by deprivation and hatred, the "Milkmen" rendered helpless by coddling and egotism.

Milkman *does* change, but he begins to change only after he moves out of his own personal present in search of history. Granted, the history he seeks is the private history of his own family—and one that he hopes will bring him personal gain in the form of the lost gold—but as it turns out, that private history is inscribed over and reflects a deeper public reality. Like Celie's father in *The Color Purple*, his grandfather's life after the civil war was typical of so many former slaves who worked hard enough to acquire property and to enjoy "the bountifulness of life." Whites, threatened by the new economic prosperity of blacks, often resorted to violence to take away their property or to drive them out of business. Milkman finally realizes that "owning, building, acquiring" was his father's "future, his present, and all the history he knew" and that his distortion of life, "for the sake of gain, was a measure of his loss at his father's death" (300). In the end, the narrator explains both Macon Dead's greed and Guitar's vengeance as reactions to personal losses, which are themselves the consequences of publicly sanctioned racist practices. But in terms of this novel, to understand is not to condone or to rest easy in the explanations found in personal history. At the end of the novel, Milkman has just begun to make the first tentative steps toward a life that reaches beyond the self to know another history. Perhaps it is in the sense that Milkman has given up the burden of self that he is able in the end to make the leap, "to fly."

In that final scene of the novel, as soon as Pilate and Milkman bury the bones of her father on top of Solomon's Leap, Guitar, who has been waiting below, fires a shot that kills Pilate. Milkman, a changed man, stands up and challenges Guitar with the words, "You want my life?" (337). Guitar, putting down his rifle, stands up, and Milkman challenges him again to take his life and then leaps into the air. With that leap the novel ends. Does Milkman fly away to Africa, leaving the vengeful Guitar on the ground? Does Guitar kill him? Does he kill Guitar? Here Morrison does not seem to leave a clue.

The ambiguity, however, is in this case an important part of the novel's judgment of what Guitar calls "the condition our condition is in" (222). The struggle within the black community between what might be called the traditions of African folklore and religion, and the traditions of American rationalism and Christianity, reflected in a wide cross-section of black literature, was articulated at least as early as 1903 by W. E. B. Du Bois.[16] If Milkman flies, then the African prevails; if he falls and dies, then the American tradition prevails. Of course, neither alternative is desirable. The final scene of *Song of Solomon* is emblematic of the ambiguous "condition" of blacks in 1963. After the short-lived harmony of the historic march of August 1963, the factions began to emerge. At the same time that Milkman is being stalked by Guitar—late summer 1963—the historical

Martin Luther King, Jr., was being stalked, politically and philosophically, by Malcolm X. King was already in conflict with the leaders of SNCC; Malcolm X would in the same year be suspended from the Muslim movement by its leader Elijah Muhammad.

Rather than winding up or finishing her story, Morrison chose to suspend it, to leave her characters poised, one in mid-air, the other putting down his rifle and standing up. The final tableau, a visual image of possibility and interdependence, suggests that to succeed all factions must somehow transcend a preoccupation with self and the personal, to become part of the greater history that affirms and loves "life, life, life, life," in order to escape the potentially fatal embrace of each others' "killing arms" (337). The condition of black Americans in 1963, the novel suggests, had two possible futures: on the one hand, for those who live on the stage of history, a future in which self is subordinated to community; for those whose lives continue to be directed by old and festering personal wounds, a future informed by anger, self-absorption, and violence.

The underlying history of *Song of Solomon*, however, does more than just suggest the "condition" of warring forces creating the alienation and fragmentation that permeated the private lives and the public movements of black Americans in 1963. By associating Milkman and Guitar with the two extremes of the black world—Martin Luther King's nonviolent Christian Church and Malcolm X's militant Muslim Mosque—Morrison incorporates into her story the entire spectrum of black experience, and at least in the fictional, nonhistorical surface of the novel, she unites the opposing forces in a kind of narrative coalition. These lives are linked by their common heritage and their interdependence. Pilate may have contrived to bring Milkman into Guitar's hands, or she may have been used and eliminated by Guitar. Guitar may kill Milkman, be killed by him, or each may transform their antagonisms into the suggested embrace of the ending. Though they were in open conflict on many fundamental issues in 1963, both King and Malcolm X were moving toward attitudes that might have made possible the union of masses of African Americans. The possible scenarios for Milkman and Guitar in 1963 are the same as those for the larger black community.

When *Song of Solomon* was published in September 1977, for many the civil rights movement was already history. The uproar following the last years of the Vietnam war and the outrage over Watergate had pushed civil rights issues off the front page of the newspapers and out of the minds of many of even the most public-minded citizens. King had been dead nine years; Malcolm X twelve years. Ten years had passed since the Detroit riots that were generally considered to be the worst of all; twelve years since the passage of the Voting Rights Act; seventeen years since the sit-ins in Greensboro, North Carolina; twenty-two years since the Montgomery bus boycott; twenty-three years since the Supreme Court's decision in *Brown v. Board of Education* declared segregation in the public schools unconstitutional. The movement had had its day.

The thirty-two-and-a-half years that make up the fictional life of Macon Dead III, called Milkman, were years of continual struggle for racial justice, which would not receive official federal sanction until the passage of the Civil Rights Acts of 1964 and 1965, at least in part as a consequence of the assassination—a few weeks after Milkman's leap—of John Kennedy, the man many black Americans looked to as their "White Hope." Black Americans would continue to pay the price for that progress, not only in the deaths of their leaders, Martin Luther King, Jr., and Malcolm X, but in personal sacrifices of the ordinary people who worked to foster racial justice.

Readers of *Song of Solomon* in the 1990s stand in a similar relation to the future as did those in the late 1970s. They have a choice of participating in the creation of or withdrawing from the public, historical text. The success of this new kind of historical novel, in which the Kings and Xs do not walk through the fictional text but remain in the undertext of the palimpsest, is in part dependent on the quality of its readers. For those who choose not to decipher the palimpsest, it may remain a highly fanciful, compelling story of personal lives, informed by the imagination and the folk myths that carry such narratives. Those who participate in creating the historical text beneath the fictional text write themselves into the crossing of history and the fiction of the self and in the process affirm the inevitable historical subtext of all our lives.

The Lakestown Rebellion

A rollicking, largely comic novel, Kristin Hunter's *The Lakestown Rebellion*, like *Sula*, focuses on an all-black settlement. Whereas in Morrison's novel the spontaneous, violent, and for many fatal, attack on the tunnel and on the white establishment it represents is the beginning of the end of a community, the carefully orchestrated nonviolent rebellion in Hunter's novel results in a joyous victory and revives and invigorates a black town whose leaders had been slowly capitulating to the demands of whites. Mired in the years 1919–41 when there was little or no promise of social change, Morrison's characters in *Sula* do not consider the possibility of combating the white establishment or, for that matter, of playing any role in the public arena. The characters in *Lakestown*, however, though far removed from the public events being played out nightly on national television in the summer of 1965, immediately counter a threat to their town with an effective plan that utilizes the proven methods of the movement. While warning against the temptation for successful blacks to collaborate with the schemes of the white establishment at the expense of the black community, *Lakestown* also celebrates those who carry on the fight at the local level after the national battles have been won.

Like *Song of Solomon*, *Lakestown* has a large cast of characters spanning the social spectrum: middle-class matrons, eccentric spinsters, unemployed men, prostitutes, maids, preachers, militants and pacifists, a doctor and a lawyer. The mayor of the town, Abe Lakes, like Macon Dead II, is descended from black and Indian blood and is married to a light-skinned woman with whom he no longer has a sexual relationship. He sees himself as superior to his fellow townsmen, whom he exploits. And, also like Macon, he has cut off his past and his family's history. Politically ambitious, Lakes has struck a deal with the white power bosses in hopes of getting a job in state government and dreams of being accepted in the white world of "silently air conditioned rooms with no odors" (12). Bella, his earthy, sensuous, often disheveled wife, takes comfort in her garden and in drink; in an affair with Abe's brother Ikie, a successful sculptor; and in the company of Abe's illegitimate daughter, Lucinda (Cindy) Coddums. Cindy's mother, Vinnie Coddums, who works as a domestic servant for one of the white men who has manipulated Abe into cooperating in their plan to destroy much of the town, initiates the rebellion that forms the major action of the novel. Fess Roaney, "a pensioned-off Navy veteran," who suffers from multiple complications of diabetes, serves first as an objective observer of the action, then as unofficial presiding officer of the rebellion. Doc Thompson, Fess's physician and best friend, is a co-conspirator. When Fess wonders what ever happened to the "spirit" of the town's founders, "the first runaway slaves, the fight that was in them, the determination," Doc responds optimistically that the dormant spirit needs "something to shock it awake" (38).

While *Lakestown* lacks the complexity and ambiguity of *Solomon*, the progress of the narratives of both novels is based on the reconnection of past and present, as characters rediscover or re-enact their past or their family's history. The major action of *Solomon*—more than two-thirds of the novel—is set in the summer of 1963; that of *Lakestown* takes place in July and August of 1965. Unlike Morrison's novel, in which events of the civil rights movement are woven into the text, the culminating episodes of the movement that are making national headlines during the same weeks that Lakestowners are carrying on their local struggle are never even mentioned. Although *Solomon* focuses primarily on private lives, the presence of the historical and public subtext is compelling and integral to the narrative.

Hunter sets her novel in history by referring to episodes in the movement preceding the novel's action, but even these episodes are treated more in comic than heroic terms. Three characters in *Lakestown* have been involved in the movement. Lukey Hawkins, a college student, inadvertently becomes part of a freedom ride and is arrested and then treated as a hero by unknowing white liberals; Ikie Lakes goes to jail for transporting "some dudes [who happened to be armed] to Martin's March on Washington" (112); Reverend Bream, who "is all for marching, singing, and praying, with a legal permit for every demonstration" has led a boycott against a local chain store that "resulted in the hiring of a single black employee" (40–41).

Lakestown is, nevertheless, a celebration of the civil rights movement in that the principles of nonviolence overcome the forces threatening the community. The characters muster their forces for a fight, create a coalition of different segments of the community to defeat the white power structure, and employ strategies that have prevailed in the movement. What works in Lakestown is what has worked in the movement. The overnight rebels seem to have absorbed their skills from the culture at large rather than from active participation in the movement itself.

Although she has spent the years of the movement doing domestic work to support her handicapped child, Vinnie Coddums has embraced the idea of both coalition and reform and serves as the catalyst for the local protest. Like Dorine in *A Measure of Time*, who hangs pictures of the antagonists Booker T. Washington and W. E. B. Du Bois—primarily because of her personal connection to the two men—Vinnie has pictures of Marcus Garvey and Martin Luther King, Jr., on her wall presumably out of respect for their activism. Both photographs eventually become weapons in Vinnie's own fight against those who exploit her as she hurls one picture and her daughter throws the other to "hasten the departure" of first a bill collector and then the police officer sent to evict her. When

Vinnie finally goes to war against the enemies of the community, she combines the principles of passive resistance with the threat of violence, if not violent acts. The remaining thirty-odd characters, however, give no indication that they even know what is happening in Alabama, Mississippi, Washington, and Watts—or that they have ever heard of Marcus Garvey. By treating the efforts of even the "activist" Lakestownians with a comic, dismissive tone, the novel suggests that an immediate, personal threat is necessary to override complacency and thrust them into public action.

The crisis confronting Lakestown is a proposed eight-lane highway that Abe Lakes had been led to believe would pass through the town at ground level, creating new businesses and bringing in money from travelers who stop for food and lodging. At first unconcerned that the highway will require the condemnation or relocation of forty-two homes and a church (29), Lakes later learns that the white men have lied to him and that the real plan is to build the highway in a depressed gorge that will destroy much more property while providing no access ramps to the town. Even after he learns of the double cross, Lakes agrees to go along with the scheme by keeping the changed plans secret so that he will be able to use the white men for his own personal advancement.

As the national struggle for civil rights legislation is reaching its climactic moment with the passage of the Voting Rights Act, Lakestownians, lulled by relative prosperity, live for themselves and enjoy what comforts they can—until Vinnie Coddums sounds the alarm and alerts the town to Abe's betrayal. At last they join in the fight, but their rebellion is as much against middle-class black collaboration with the white power structure as it is against racism. While the battle for integration of public facilities and for voting rights was being won, the novel seems to suggest, old patterns of exploitation were continuing and would have to be fought on the local level.

An imaginary town in New Jersey, Lakestown is no ordinary community, for though it may not be making history in the present it has a distinguished past. Before emancipation, it was the first stop on the Underground Railway. Its original settlers were runaway slaves who "risked cruel punishment or death . . . to flee slavery in South Carolina" and who, once they settled in New Jersey, continued to risk their lives as they aided other runaways and battled "bounty hunters" (21). In 1929 it seceded from a nearby township to become "free, black, and independent" (21). But by 1965, the spirit that created this special town seems to have died: "The fires of revolution were raging everywhere else, but Lakestown would be the last place in America to ignite" (20). According to one character, the spirit that built the town has been "put to sleep by comfort and prosperity" (38).

The discovery that the community is threatened turns out to be the "something" that Doc had hoped would shock the inhabitants of Lakestown and unite its disparate and sometimes conflicting factions in a common effort to suppress both the external and internal forces that seek to destroy it. The rekindling of the spirt of rebellion is also a reconnecting of the present with the past, a process that the novel consistently affirms. Various characters discuss the value of history and how the past informs and transforms the present.

Not all the characters find history valuable or positive, however. Ikie Lakes, who claims to have "learned history" from both his Indian grandmother and his black parents, argues that history teaches resignation, that the white man, who has simply "loaned" the land to Lakestown inhabitants, creating a "fancy labor camp," takes what he wants when he wants it and "there's not a damn thing we can do about it" (88). But Doc counters that the town is "*owned*" by blacks and that under "almost any backyard" are the bones of slave owners who came to recapture runaway slaves (88). When Ikie dismisses the validity of such stories and calls Fess "boy," Fess retaliates in a rage: "Don't call me 'boy.' I'm older than you. I know more history than you, too. And I can tell you, it does help. It can give people determination and courage" (89).

Fess's voice here and elsewhere in the novel has the ring of authority, and when he goes about convincing the city councilmen to attend the meeting where the news of the highway's route will be divulged, he appeals to their pride in "Lakestown's history" (145). At the meeting of the Lakestown Borough Council, some argue against the highway on the grounds that Lakestown is an important historical community—"the first northeastern stop on the Underground Railroad"—with houses, a church, and a burial ground of historical importance (164–65). The townspeople are convinced to participate to defend both their homes and their history.

Within hours after the council meeting, secret plans are underway to stop the highway. Just as civil rights activists had done, the conspirators congregate in pool halls, turn church services into strategy meetings, and stir their spirits singing movement songs. Lukey Hawkins, the reluctant hero of the freedom rides who found it to be "an awful strain, being a hero" (19), is transformed into an effective leader of the Lakestown rebellion. Once the rebels are at work diverting the attention of the highway workers and thwarting the progress of the road in every sneaky way they can think of, Fess insists that the forces of history have been awakened and that the apathetic present has been reconnected to the past: "You know what we're watching, Lakes? A reenactment of history. Think of the organization it took to escape from slavery. Think of all the strategies and tricks it took to move all those people. A network stretching for hundreds of miles and ending right here. And now we're starting it up all over again" (247). The sense of history that feeds the determination and courage of the rebels also nurtures a sense of common purpose and unity.

The battle against the white power structure is fought on many fronts by a coalition of religious leaders, community leaders, church women, children, and ordinary working men and women, as well as by those on the fringes of society—tavern owners, two prostitutes, and a gang of teenagers called the Young Warriors. As she and others are planning to halt the highway project, Bella Lakes warns that "black people . . . have to work together as a tribe. We're surrounded and outnumbered. . . . this is something that could involve every man, woman, and child in town" (90). On the night before the rebellion begins, Ikie Lakes calls for "a united effort" (222), and for almost two months Lakestownians use every trick conceivable to disrupt the progress of the highway construction: in a single night, they fill in a large excavation that took days to dig; children offer the workmen drugged lemonade, the women tainted food; church people set up a revival tent and proselytize to the confused workmen, while prostitutes come along to tempt them away from their jobs. After dark the men sabotage the expensive but delicate earth-moving equipment.

In the end, when the workmen have created a "desolate dust bowl" that "seemed vast as the Mojave Desert, bottomless as the Grand Canyon" (283), Fess and others conspire to dam Crump's Creek and redirect the flow of water into the cavern. After drugging the guard and tranquilizing the guard dogs, the men set to work. With the help of three days of unrelenting rain, the ugly construction crater is transformed into "a shimmering sapphire mirror at least a mile wide," and Lakestown has a lake at last (306).

This largely comic story includes tragic elements: an outcast boy accidentally drowns another boy after they invade the swimming pool of a white country club; Lukey Hawkins loses a foot while driving the earth-moving equipment; and Fess Roaney drowns when the dammed-up water is released into the construction crater. But the comic mode dominates in the end. Abe's illegitimate child, Cindy, who with the exception of her ability to sing, has been mute all her life, learns to talk; when Cindy gets pregnant, Vinnie Coddums and Abe's wife, Bella, preside over what starts as a shotgun wedding and ends with Abe stepping in to give away this daughter he has never acknowledged. In the final scene, Abe declares "Black Independence Day in Lakestown" (312), Ikie christens the new body of water "Lake Lucinda" (308), Bella impulsively strips and joins the swimmers cavorting in the water, and Abe at last accepts his wife and himself.

The comic mode and spirit of celebration reinforce the novel's basic optimism and Hunter's commitment to creating narratives that show that problems can be solved and to dramatizing "the enormous and varied adaptations of black people to the distorting, terrifying restrictions of society."[17] Published in 1978, *Lakestown*, like *Song of Solomon* the year before, speaks into a time when few people, black or white, were optimistic about realizing the dreams of racial harmony and a just society that characterized the last years of the movement. Since 1965 Americans had watched the erosion of hope for real social reform, as domestic riots destroyed large sections of our inner cities, economic and human resources were diverted to the Vietnam War, and many became increasingly cynical after the revelations of Watergate. Jimmy Carter's commitment to human rights and peace in the Middle East restored some hope for a more humane public agenda, but by the time *Lakestown* was published, there was little reason to expect that he would solve the mounting domestic problems that continued to take the greatest toll on the urban poor, many of them black.

For those who felt that a milestone had been passed with the Voting Rights Act in August 1965, the events of the next thirteen years were anything but fulfillment of earlier dreams. The assassination of King, 150 major urban riots, the continued disintegration of alliances within the black community, the election and re-election of Richard Nixon, Jimmy Carter's failure to counter the prevailing forces of reaction—these and other discouraging events had created a sense of despair about the future. Just as Vinnie Coddums has alerted her fellow townspeople to the immediate threat to their community, *The Lakestown Rebellion* is a wake-up call for those who have lost hope for further progress—perhaps for middle-class readers who have given up the struggle and, like Bella and others at the beginning of the novel, are taking their comfort where they can while those in power continue to undermine the integrity of the black community.

Unlike *Solomon*, *Lakestown* smooths over the conflicts between private and public commitments. In spite of years of hostility, Bella and Abe suddenly overcome their differences by joining in a common struggle. Abe's conversion, seemingly as miraculous as that of Mister in *The Color Purple*, implies that good politics makes for good relationships—even good bed-fellows—and vice versa. In spite of deaths and injuries, *Lakestown* ends as a comedy that provides easy resolutions to what Morrison sees as inevitable conflicts between private and public lives.

To the readers, then, especially disillusioned, middle-class readers in 1978, the novel warns that African Americans cannot trust either the white power structure, the federal government (represented by the highway commissioners), or blacks who benefit from cooperating with the white world at the expense of the black community, like Abe Lakes before his transformation. The success of rebellion suggests that blacks would do better by using their collective wits against whites than by resorting to violence or waiting for help from established institutions. Those with vision and a sense of history, like Fess Roaney, may have to persuade their neighbors to act: black men and women must remember their heroic past and reconcile their differences—as Abe and Bella do—and become allies in the struggle to build better lives for themselves; all must wake up to the need to cooperate for common goals.

Perhaps the only novel considered in this book that suggests a positive program, *The Lakestown Rebellion* affirms public but local activism, without the hoopla of television cameras. Further, its characters who work for the common good find their private battles resolved in the process. Bella and Abe restore their marriage not by privately talking out their considerable differences but by joining a cause that is greater than they are; in the process, they transform their private lives.

Like *Song of Solomon*, *Lakestown* portrays the black community threatened by internal conflicts. But while Morrison's novel explores the tenacity of egoism and its power to isolate individuals from each other and block their participation in communal action, Hunter celebrates a fictional black community that unites to defend against a common enemy. Hunter recognizes the distinction between optimism about society and the happy-ever-after world of her fiction: "One of my motivating forces has been to recreate the world I know into a world I wish I could be in. Hence my optimism and happy endings. But I've never dreamed I could actually reshape the real world."[18]

Sassafrass, Cypress, and Indigo

The focus of *The Lakestown Rebellion* is public and communal action, and the conflicts of the characters' personal lives are resolved through their participation in that action, while in *Song of Solomon* the protagonist's prevailing egoism and preoccupation with the personal is held in narrative tension with the compelling public events. Ntozake Shange's *Sassafrass, Cypress, and Indigo,* set in the same period, is insistently personal, its characters preoccupied with their relationships, their art, and their feelings—with and without drugs. The one character who is raising money for the movement seems to be motivated not by political action but by the man in her life. *And* she does not anticipate that there will be any conflict between a public commitment and her private life. Like *Lakestown,* then, *Sassafrass* assumes that there are no inevitable conflicts between the two nor any necessity of making difficult choices. Both novels seem to suggest that, in the parlance of the late seventies to early eighties, "you can have it all." Yet neither Shange nor Hunter is inordinately sanguine about the possibilities of living fully in our times.

Sassafrass seems to sanction the self-expression, idiosyncrasy, and individualism associated with the doing-your-own-thing sixties. Composed of fragments—bits of folklore, letters, poems, dreams, recipes, journal entries, and passages of sustained narrative—this unusual modernist novel relates events in the lives of the main characters, Sassafrass, Cypress, and Indigo, the three daughters of Hilda Effania, a widow in Charleston, South Carolina. Since her husband Alfred died at sea, Hilda Effania supports her family by weaving exquisite fabrics and making clothes for wealthy white clients. All three girls are craftswomen or artists. As a small child, Indigo makes dolls from ribbons, buttons, and left-over scraps of fabric. When her mother insists that she give up these dolls, who are part of an elaborate fantasy world she has created, Indigo learns to play the violin with such feeling that she is called upon to bring relief "from elusive disquiet, hungers of the soul" (222). Sassafrass, a poet and also a weaver like her mother, thinks of her cloth making as an activity that connects her with women "from all time and all places" (92). As a dancer Cypress learns to use her body to create an art that is at once beautiful and political, a dance that is "the essence of the struggle of colored Americans to survive their enslavement" (136).

Unlike the precise, complex dating of *Song of Solomon* or the specific temporal setting for the main action of *The Lakestown Rebellion,* the dating of this novel seems casual and haphazard. The only specific dates in the novel are 1946, the year Hilda Effania and Alfred were married; reference to a 1959 car; and a song lyric that includes the words "circa 1963" (88). The girls' ages suggest that the events of the novel occur after 1963. Any effort to identify the dates more precisely leads to inconsistencies and apparent anachronisms. For example, when Cypress announces that she is joining a dance troupe to help "raise money . . . bail, legal fees . . . for the Civil Rights Movement" (210), she refers in the same part to the bombing of a church—suggesting Birmingham in 1963—and to Rap Brown and the Panther party, prominent only after 1966. There are also references to the war, presumably Vietnam, and to the moon, where "white people put flags and jumped up and down" (6), which of course did not happen until 1969.

Published the same year as *The Color Purple*, *Sassafrass* slides over historical realities in much the same way. Like Celie, who builds a successful business making pants for women at a time when few women were wearing pants, Hilda makes enough money as a weaver and a seamstress to maintain a middle-class household and to buy luxuries for her children in a Southern city where such material success for black women was unlikely. Also like *The Color Purple*, this unusual novel validates those who seek their own pleasure at the expense of others. In both novels characters take up and discard lovers with little regard for the emotional consequences, and Shange's characters emerge apparently unscathed from drug and alcohol binges, as well as from reckless sexual escapades. Further, *Sassafrass* does not judge the three young women, who, except for very occasional visits, do not include their adoring mother in their lives.

Hilda Effania's ambition for her talented daughters is quite different from what they want for themselves: "Hilda wisht her husband Alfred could see the girls lined up by the kitchen sink, taking a ballet lesson from Cypress, while Sassafrass recited Dunbar. They were so much his children: hard-headed, adventurous, dreamers. Hilda Effania had some dreams of her own. Not so much to change the world, but to change her daughters' lives" (57). When she is alone, she talks to Alfred, explaining how she is raising their children, justifying her ambitions: "Yes, Alfred. I think I'm doing right by 'em. Sassafrass is in that fine school with rich white children. Cypress is studying classical ballet with Effie in New York City. . . . I'm sending Indigo out to Difuskie [*sic*] with Aunt Haydee" (72).

Hilda, who must cater to the whims of frivolous white women, expects her daughters to be free from such obligations, but she does not consider that it might be necessary to change the world to make that possible. In fact, she talks to her dead husband and tells him that she "can't imagine another world" (72). She sends her youngest child, Indigo, to nearby Daufuskie Island in part to avoid "all this violence 'bout the white & colored going to school together, the integration" (72). Hilda does imagine, however, a private happiness created as her girls find "Nice husbands. Big houses. Children. Trips to Paris & London" (72)—in short, the rather closed world of the black bourgeoisie that would have been out of the reach of most black women with Southern origins struggling to succeed in the world of dance, art, and poetry.

When Sassafrass and Cypress first leave home what they actually find is the opposite of their mother's dreams of nice husbands, big houses, and trips to Europe. Sassafrass is living in Los Angeles with Mitch, a not-very-nice man, an ex-convict and sometime junkie, who beats her. Far from longing for the antiseptic world of "nice" people that her mother writes her about, she looks forward to "the black revolution" or to the time when she and Mitch will have enough money to move to a "black artists' and craftsmen's commune . . . outside New Orleans . . . near a black nationalist settlement" (77). But her mother, who objects to her daughter's plan to "take the white folks to court to get back land they've owned since before the war," chides her for attacking "*the white folks in the middle of one of their wars*"—presumably Vietnam (131). But Sassafrass's intentions to become an activist have little basis in reality. Once in the commune, Sassafrass finds not so much a political role but sexual indulgence, pregnancy, and something she thinks of as "spiritual redemption" (213).

Sassafrass's flirtation with black nationalism is never fulfilled in the novel, nor is the issue of separatism resolved in the narrative. When she finds that she is going to have a baby—fathered either by Mitch or by her guru, Shango, in a ritual celebration—she decides to return to her mother, who is convinced that her children have gone wrong because "the world's going crazy and trying to take my children with it" (220).

Cypress begins her career when her mother sends her to New York to study dance. There she joins a troupe of black dancers dedicated to "discovering the movements of the colored people that had been lost" (136). Later, she lives in San Francisco—dancing, sleeping around, snorting cocaine, and dealing dope when she needs money. Back in New York, she gives up drugs, joins a women's dance collective, has a brief affair with one of the dancers, and finally falls in love with Leroy, a serious musician and an old friend, who comes from "a long line of civil rights activist doctors and preachers" (186–87). To be with Leroy is to join her life to one who is committed to finding and preserving "the blood of the culture, the songs folks sing, how they move, what they look at, the rhythms of their speech" (189); to dance is to cling "to her body, the body of a dancer; the chart of her recklessness, her last weapon, her perimeters: blood, muscle, and the will to simply change the world" (208).

The final parts of the novel relate the outcome of the three sisters' adventures. Cypress has a socially significant career, as well as a satisfactory personal life: she plans to join a dance company called Soil & Soul that travels around the country raising "money and morale . . . bail, legal fees, stuff like that for the Civil Rights Movement" (210); she intends to marry Leroy, whose middle-class parents were killed by white gangsters but who is determined to nurture and create black culture and never to succumb to hatred and revenge. Absorbed in the folklore of Daufuskie Island and in practicing magic, Indigo helps preserve the folk practices of her ancestors. In the final fragment, Sassafrass is giving birth, with Indigo serving as midwife, Cypress encouraging her with the conviction that the baby will be "a free child," and Hilda comforting and accepting her wayward daughter (225). By ending this novel with one sister living and working within the African-American folk culture, another making a serious commitment to art, marriage, and the civil rights movement, and a third giving birth, Shange seems to affirm all three choices. Some women, like Indigo, will be caretakers and preservers of the past; others, like Cypress, will create an art that is consistent with, rather than in conflict with, the good of the race; still others, like Sassafrass, will follow traditional women's roles, weaving cloth and parenting—without the help of a man—those free children that will make the future.

In the final scene the four women are united through the narrative and their mutual task of bringing a new baby into the world, and at least for that moment, they function as one. None of the four main female characters—mother and three daughters—seems to be a three-dimensional woman. They are types, or aspects of woman's experience; together they form a human collage. By creating a novel without an authoritative narrative voice that makes judgments about the characters, Shange entices her readers to judge and reach their own conclusions about these women's choices. Most readers would probably conclude that Sassafrass is better off without the battering Mitch, that Cypress will be happier without drugs and married to Leroy, and that Indigo's commitment to preserving the folk culture is to be honored. Like *The Color Purple*, this novel has a happy-ever-after sense of closure. It does not, however, resolve the public issues relating to drugs, politics, and sex that the characters encounter.

The substance of this novel is finally overwhelmingly personal; yet slipped into the narrative—though never integral to it—are nods to racism, civil rights, and activism. Hilda Effania explains that her long-dead husband went to sea because "there wasn't much work for a skilled Negro carpenter in Charleston" (190) and because as a dark-skinned black from the South Carolina islands he also suffered from color prejudice in the eyes of other blacks, who were hard "on darker members of the race" (190). Though she endures the patronage of white women, Hilda is not portrayed as a victim, and she encourages her children not to fight racial injustice but to "Let it lie" (190). Although Cypress eventually finds work that is compatible with Leroy's commitment to "Blackening up America" (189), she is deeply involved with him before she knows anything about his background or his political views. His politics, like his inherited wealth, are an added attraction to this man she already loves. Although the novel implies that there is political significance to the characters' work, in the end we know far more about their anatomy, sexual preferences, tastes in food, drug use, and drinking habits than we do about their political ideals or social values. Large public issues that continue to plague the black community are raised and finally dismissed from the narrative.

When *Sassafrass, Cypress, and Indigo* was published in 1982, the majority of black women in America were—as they are today—living in poverty. Relatively few had the resources to give their children the privileges and luxuries that Hilda provides for her talented daughters. What this novel about a widowed mother and her children has to say about life in the eighties, at a time when many of the gains of the civil rights movement were being threatened, seems to be that talented resourceful women with nurturing, understanding mothers have many options, regardless of race. The fact that none of the characters specifically suffers from living in a racist society, while both Sassafrass and Cypress are victimized by men, makes the text more at home in the feminist than the race-conscious domain. Very little would change if references to race were removed and Cypress had joined the antinuclear movement instead of the civil rights movement. In fact, if that had been the case, her long dream about a nuclear holocaust would make more sense.

Like *The Lakestown Rebellion* this novel is a celebration, and like Shange's choreopoem, *For colored girls who have considered suicide when the rainbow is enuf,* it is an affirmation of black women's courage and ability to prevail under very difficult circumstances. The final episode when Sassafrass has come home to have a baby, Cypress to get married, and Indigo to help has a positive tone and suggests that all will be well. "Mama was there," and Mama is Hilda Effania, who has encouraged her daughters to seek their own pleasure in the world and not to worry about the past. *Sassafrass* may well have bolstered the morale of middle-class readers in the Reagan years, who were primarily concerned with finding satisfying personal relationships and fulfilling work, and perhaps, like Cypress, the added extra of believing that their work is socially significant—the eighties dream of having it all.

Like *The Lakestown Rebellion*, Shange's novel *Sassafrass, Cypress, and Indigo* glosses over the conflicts that Morrison sees as inevitable between private and public commitments. Like *The Color Purple*, it implies that major changes in human character and behavior are accomplished with apparent ease, as characters kick drugs, change lovers, and find meaningful work without significant cost. At a time when those in power were denying the reality of racism and suggesting that the drug problem would be solved if we would "just say no," this novel of easy answers provides the consolation that so many sought and some continue to seek.

Betsey Brown

Shange's *Betsey Brown*, on the surface, concerns the growth of the moral and social conscience of twelve-year-old Betsey in the context of the turbulent activity of her large middle-class household. The household consists of her parents, Jane and Greer Brown; their other children, Margot, Sharon, and Allard; their adopted nephew, Charlie; and Jane's aging mother, Vida Murray. The time is 1959, the place St. Louis, Missouri. Betsey is an especially privileged child. Her handsome doctor father and beautiful, glamorous social-worker mother, more than happily married, live in a very large, attractive Victorian house. The children, who occasionally squabble over petty matters, enjoy each other's company and do their chores with surprising good cheer. Betsey, the oldest, is smart and pretty; she even has a boyfriend who seems satisfied with kisses. This generally happy family does, however, have problems, most of which are related to, if not caused by, racism, but the tone of the novel and the way its conflicts are resolved are more consistent with an episode of The Bill Cosby Show than with most novels treating the impact of public events on private lives in the African-American community.

Trouble-making racism begins at home, as Vida, who thinks of herself as "most white," insists that "slaves and alla that had nothing to do with her family, until Jane insisted on bringing this Greer into the family" (19). She objects to her son-in-law because he is "dark and kinky-headed" (18). Grandmother Vida not only worries about the African characteristics that her grandchildren have inherited from their "jet black" father, but she opposes Greer's involvement in the civil rights movement, believing that it is "best to be the best in the colored world, and leave the white folks to their wanton ways" (30). While Greer can make light of Vida's color and class prejudice and indulge her with little gifts and good-natured teasing, he finds that racism from outside affects his family in a far more serious way.

Until *Brown v. Board of Education* forces them outside their own community into the white world, the children have been relatively sheltered by living in a middle-class black neighborhood with "Negro-owned businesses" (91). They have, however, encountered racist behavior and attitudes before the busing order. One day, Betsey and her friend Veejay visit their white classmate, but the girl urges them to leave before her mother returns since "weren't no 'niggahs' 'sposed to be in the house" (41); that same day, Charlie and Allard get in trouble for riding their bikes on the grounds of a Catholic school where "colored weren't allowed to do anything . . . not even the cleaning" (45). When "two huge red-faced white policemen" warn Jane that she must not allow her children to trespass on "private"—that is, white—property, Jane thanks God that "they hadn't whistled at some girl like poor Emmet [*sic*] Till murdered in Mississippi only four years before" (45).

Although the narrator observes that Greer and Jane have both "been chastened since birth by the scorn and violence the race had known . . . [and] brought up on lynchings and riots, namecalling and 'No Colored Allowed,'" they reach conflicting conclusions about how to deal with racism. Jane, afraid of whites and hesitant to expose her children to them, chooses to keep "her world as colored as she could" (90). Like Hilda Effania in *Sassafrass, Cypress, and Indigo* she believes in protecting her private world and leaving the race issue alone. Greer, on the other hand, takes action to change the conditions that allow racism to persist and actively works to alleviate the suffering of the economically disadvantaged.

While Jane has a cavalier, even heartless, attitude about underprivileged blacks, Greer goes out of his way to care for the poor—making housecalls and helping them find jobs. When Greer hires Regina Johnson to help with the children, he takes her on as a "special project":

> Every one of the Johnsons had something wrong with them. Something that came from too little of everything. Not enough food, not enough exercise, not enough light, not enough love. Got to the point that old man Johnson most gave up, that's why Greer invested so much time and energy with them. He couldn't stand the idea of losing another colored family to the pressure, not just high blood pressure but the pressure of little rooms smelling of too many people and little wants feed-

ing big hungers for light and air. Stairs had to smell of more than oldness and urine. (82–83)

But Jane thinks of Regina as a convenience, like her washing machine, rather than a person with needs and a social context of her own. When Regina quits her job, Jane does not inquire about the circumstances but rather complains that she has left at a particularly stressful time.

Twice in the novel, Jane fires domestic servants without a thought for their welfare. Early in the novel, when she comes home and finds the house in disarray, she dismisses poor, uneducated Bernice without bothering to find out that, in fact, the children are responsible for the mess. Later, when Carrie, an older black woman who has cared for the children while Jane is away, calls from the jail to explain that she has been detained and cannot come to work, Jane immediately dismisses her, callously demanding that she take her "mess right out of this house," without even inquiring why Carrie found it necessary to "cut a friend" (205–06).

An order to integrate the public schools by busing black children out of their neighborhood into white schools creates havoc in the seemingly idyllic Brown household. Thoughts of the lynching of Emmett Till plague the Browns on the day the children take four different buses to widely scattered, predominantly white schools. Charlie, already a teenager, frightens the children with tales of what might happen and reminds his Aunt Jane that Till was just his age when he was murdered. As they are leaving the house, terrified by "peckerwoods" and "thin-lipped rednecks" (90), Charlie makes matters worse by threatening to "get some white tail . . . for Emmett Till" (97). Charlie does come home after his first day in an integrated school with torn clothes and bruises.

A more serious threat to the family's happiness comes when Dr. Brown defies his wife and takes the children to a civil rights demonstration to protest discrimination at the "racist paragon of southern gentility, the Chase Hotel" (157). Afraid for the children's safety, Jane accuses Greer of putting his public commitments before his private responsibilities. He, on the other hand, insists that it is time "to do something about . . . this separate but equal travesty" (156) and that by teaching his children that the struggle for civil rights is their struggle, he is doing "what a man's supposed to do for his wife and children" (159). When Greer refuses to give in to her demands, Jane decides to leave him and the children: "I will let you know, if I'm coming back. Just tell the children Mommy went away for a while" (160).

Greer, nevertheless, feels compelled to go through with his plans to teach his children about public responsibility: "Greer felt the front door slam all the way in the kitchen. He sat down praying she'd understand and come back. . . . He was taking his babies to a battle he wasn't sure he'd win. He was leading his children of his own free will to face what grown Negroes had already died for" (160). The demonstration turns out to be peaceful, and Greer concludes that he did the right thing: "A man had to stand for something" (161).

In the end, Jane returns to her husband with a "new understanding of him, what he stood for, for their people, for the children" (190). There is nothing in the novel to suggest, however, how or why Jane has changed. Greer doesn't ask, and she offers "no information" about where she has been. The family celebrates "the progress of the race" along with her "homecoming," but there are no specific references to progress in civil rights, not even whether the peaceful demonstration at the Chase Hotel that precipitated the family crisis accomplished anything (190). And, there was very little progress in civil rights in 1959 to celebrate.

The conflict between Jane and Greer is resolved, then, in a personal context. The narrator specifically notes that they choose not to discuss what has happened and that their differences are dissipated in the heat of their lovemaking, their very different political views having been swept literally under the bed. In the end there is nothing in the novel to suggest that Jane will join in the movement. Her return is motivated not by a fundamental change but by thoughts of Greer: "his arms, his chest covered with twidly nappy black hairs, his hands stroking her hair from right to left, the moustache across his lips" (189). Shortly after her return, Jane is preoccupied with her children's sex education, her primarily sexual reunion with her husband, and the endless problems with her domestic help. Jane's newfound respect for her husband's ideals and for "their people" does not seem to affect her treatment of the poor elderly black woman who has cared for the children in her absence; immediately after firing her, Jane turns "from the telephone quite her personable self" (206).

But while much of this relatively short novel focuses on the private world of Jane and Greer, it is finally Betsey's story that frames the narrative and shapes the action. Perhaps influenced by her father's zeal for social reform, Betsey discovers that her carelessness can be responsible for the suffering of others after she thoughtlessly harasses the new housekeeper. When Jane fires the woman, Betsey's friend Veejay, whose own mother does domestic work, accuses her of acting "like white people" and making "things hard on the colored" (67). Betsey feels guilty: "A heavy red glow came over Betsey's body. Shame. She was ashamed of herself and her sisters and Charlie and Allard. Veejay was right. Bernice just talked funny was all. Betsey'd passed over the paper bags fulla worn-out clothes, the two shoes of the woven cotton, fraying by the toes, and the calluses on the palms of the woman's hands. Betsey Brown had been so busy seeing to herself and the skies, she'd let a woman who coulda been Veejay's mama look a fool and lose her job" (68).

Betsey determines to correct what she has done, but when she fails to get her mother even to listen to her confession, she decides to "do her best not to hurt or embarrass another Negro as long as she lived" (70) and to be different from the "nasty white children who bothered Veejay's mother" (71). Powerless to change the situation, Betsey decides "to do penance instead" by having a good cry "for having cared so little" (71). Like Harker in *Dessa Rose*, published the year after *Betsey*, she "feels bad" for those that the more clever and resourceful of her race have "left behind."

At the end of the novel, Betsey is "making decisions and discoveries about herself that would change the world" (207). Just as there is no evidence to support the implication that Jane has changed, Betsey's determination to be a world changer is not supported by specific plans or ideas about how she might reform the world. Nor are the time and place ripe for fostering Betsey's spirit of reform. The year is 1959, and Betsey is in the seventh grade; much of the work of the civil rights movement will take place in the next six years and will culminate in the Voting Rights Act, about the time Betsey reaches her eighteenth birthday. Exactly what role she will play in the movement as a privileged teenager living in St. Louis, a city that saw very little civil rights activity, is not clear. Lacking a narrative link between the time of the novel's setting and the time of publication, *Betsey Brown* leaves readers in a once-upon-a-time world when well-meaning privileged youngsters could feel good about themselves simply by being emotionally on the right side of a cause.

The novel's few passages focusing on Betsey's growing social conscience are interwoven with what to her are the much more compelling aspects of her growing up: a boy named Eugene Boyd, whose kisses are "soft and light, like petals of protea or Thai orchids. . . . a river wisp and innocent as dawn" (75); a desire to "paint her nails and wax her bangs" and then to do what the "bad girls" do (117); a fantasy about being chosen "Queen of the Negro Veiled Prophet" and riding on a float on a night when "the stars were sapphires, opals, and diamonds, a tiara for a queen" (140). While her fantasies are mixed up with notions about being "in the trenches for the race" and putting "white folks in their place," Betsey expects to do so by winning "the dance trophies" (117) or joining Tina Turner as an "Ikette" (207).

Betsey's private longings, though occasionally mixed with a vague notion about social reform, are largely consistent with her mother's values. When Betsey runs away from home, she goes to a beauty shop and has her hair and nails done, rather than seeking out the offices of a civil rights organization. After returning to her family, Betsey muses about the future: "She'd just wanted to see the world. Marry a Negro man of renown. Change the world. Use white folks' segregated restaurant tables to dance on, and tear down all the "Colored Only" and "Colored Not Allowed" signs. . . . She wanted to swing her new hairstyle and have her Humphrey Bogart not be able to keep his eyes off her, while she smuggled rifles for the Resistance" (152). In the last passages of the novel, Jane reinforces the "see the world" and ignores the "change the world" part of Betsey's fantasy. Advising her daughters about how to act like ladies and attract "nice young men," Jane insists that, if they take her advice, they will soon "be sashaying down the Champs Elysées with a handsome new husband" (194).

The novel does not sanction Jane's values and behavior, nor does it condemn them. While Betsey wonders why her mother doesn't understand that their housekeeper Carrie "wouldn't hurt nobody less they hurt her a whole lot" (207), neither she nor Greer intervenes to save Carrie's job. Again Betsey feels bad—"sadness . . . in her every sinew"—for a poor woman who has lost her job (206). After Betsey goes off to school that morning, Jane sits down to breakfast with her "beloved husband," and presumably all is once again right in *their* world.

Betsey Brown, like *Sassafrass*, offers relatively vague or easy solutions to the conflict between public and private life. The passionate physical attraction between Greer and Jane seems adequate to smooth over their fundamentally opposing attitudes toward the movement. He is committed to having a public *and* a private life; she, to a strictly private one. She leaves him, *and* she comes back. There are no deaths, no serious injuries, and no residual hard feelings. Conflicts are dissipated again and again in the heat of their lovemaking. No one is finally ever tested.

By 1985, the work of the movement had become history, and, as we have repeatedly noted, the progress that was made by the movement had been eroding for some time. The failure of school officials to integrate the schools effectively was once again an issue in St. Louis, as elsewhere. In the context of 1985, *Betsey Brown* seems to suggest that privileged readers need to be sensitive to the needs of others and take some kind of public action, but not to worry, they can do so without seriously jeopardizing their interesting, pleasure-filled lives.

CHAPTER SIX

IN THE WAKE OF THE MOVEMENT

The three novels considered in this chapter all belong to the post-movement period. Published in the five years from 1976 to 1981, they are all set primarily in the mid to late 1970s. Alice Walker's *Meridian* (1976), a novel structured from discrete fragments, begins and ends with its characters struggling to find a way to live in the middle of that troubled decade, just before the Bicentennial; Toni Cade Bambara's *The Salt Eaters* (1979), set in 1977, fills in the past of the characters' lives with flashbacks; Toni Morrison's *Tar Baby* (1981) opens in what seems to be the fall of 1979, though the specific date has to be deduced. While many of the fragments of *Meridian* relate events from the days of protests, marches, and constant jailings, the opening and concluding scenes situate the heroine in what is clearly the present—that is the middle 1970s.[1] Similarly, though many of the characters in *The Salt Eaters* recall and mull over events from the civil rights movement, the narrative returns again and again to the late 1970s. *Tar Baby*, on the other hand, moves steadily forward through the last weeks of the seventies and into the fall of the first year of the new decade, leaving the characters facing an uncertain future just as Ronald Reagan was about to be elected to the White House.

The young activists in Alice Walker's *Meridian* are in their late teens when they respond to the powerful pull of the civil rights movement. They are in their twenties at the time of the Selma marches and around thirty in the narrative present of the novel. The characters in Bambara's

The Salt Eaters were well into adulthood in the movement's heyday and middle-aged in the novel's present. Both Walker's and Bambara's activist characters reached adulthood at a time when joining the movement was intensely compelling. Though they chose to do it, that choice involved consequences that none could have foreseen. And both novels are about the consequences of choice.

In Toni Morrison's novels the civil rights movement, its prehistory, and its aftermath lurk just below the surface, but in her vision the movement and the stream of history in which it flows are primary. Both Jadine and Son in *Tar Baby* entered what Alice Walker calls the "age of choice" too late to have been active in the movement. He came along in time for Vietnam; she in time to be seduced by the materialism and self-indulgence that came in its wake. The only activist in the novel is Michael Street, who chided Jadine during college for abandoning the black culture and for not "organizing or something" (72).

While *The Bluest Eye*, published in 1970 and set in 1941, reinforces the importance of the movement in creating ways for black people to discover and express the beauty of blackness, the young African-American characters in *Tar Baby*, confident of their beauty and their abilities, are tortured by the seemingly contradictory choices available to them. Son refuses an opportunity to go to college, not because he doubts his aptitude but because he refuses to buy into the white value system. Jadine, on the other hand, is co-opted by the international beauty business. Dressed in "honey-colored silk" and adorned with "heaps of gold," her image on the cover of *Elle* projects a kind of "beauty" unavailable to most women of any race (117). Far from feeling any obligation to contribute her talent, her education, and her beauty to the African-American community, Jadine never even considers taking an active public role.

Yet, paradoxically, though she was still a child when the movement ended, Jadine looks back at those days in terms of what she did not do. Though she would have been only eleven years old at the time of the Selma marches and the passing of the Voting Rights Act, she feels guilty; this guilt and its expression in Jadine's nightmares suggest that she has choice about how she lives her life and that she is responsible for her choices.

But like *Tar Baby*, *Meridian* and *The Salt Eaters* are also about the far more difficult choices the characters must make in a social context that does not mandate decisions. In the mid and late 1970s, hardly anyone knew for certain what they might do that would make a difference. Many first readers of these novels might have longed for what Meridian calls "a time and a place in History that forced the trivial to fall away" (84).

In published interviews and public statements, Morrison has repeatedly defined "freedom" not as the capability to do as we like, but the ability to choose our responsibilities.[2] In the post-movement days, choosing public responsibilities has proved for many to be a difficult task; some, like Bambara's Ruby, are almost overwhelmed by the choices required by those seeking to be "responsible"; others, like Walker's Anne-Marion, escape into very early retirement from activism; still others, like Morrison's Jadine, have opted out of the responsibility business altogether. All three of these novels, composed in the post-movement, pre-Reagan years, end with characters in the process of making choices. Both *Meridian* and *Tar Baby* leave characters taking a course of action that a moment's wavering could reverse. *The Salt Eaters*, on the other hand, ends as a catastrophic, perhaps apocalyptic, event forces characters to break old patterns and make new choices about what being socially responsible means.

In the wake of the civil rights movement, during the years in which these novels were written, surviving activists were all struggling to define available choices and to identify the next steps in the seemingly endless struggle for racial justice.

Meridian

Alice Walker's *Meridian* directly concerns the conflicts between personal needs and public commitments of young civil rights activists. Although they experience considerable ambivalence about how to deal with their conflicting demands and desires, and while the novel explores the consequences of their inevitably inconsistent choices, the book is unequivocal in its presentation of young activists paying an enormous personal price for their public actions. Scene after scene dramatizes their physical and emotional exhaustion, rending breaks with family, and the devastating consequences of the sexual politics that invaded the movement in its last years. As one who lived through the movement and experienced many of its battles firsthand, Walker presents a picture quite different from Shange's sanitized, indirect, off-stage version.

In *Betsey Brown*, children are roughed up, only to rally and go about their business; *Betsey Brown*'s one dreaded demonstration turns out to be free of violence. In *Meridian*, however, heads are bloodied—and brains damaged—women raped, children jailed, families torn apart, lives irreparably devastated. But there are no scenes showing a united community euphoric with a sense of common mission; despite its action being embedded in the movement, conspicuously absent are references to particular episodes such as the 1963 March on Washington, the Birmingham campaign, Mississippi Summer, the Selma march, or the incidents of urban violence that followed the passage of the Voting Rights Act. While no one refers to specific civil rights organizations, the experiences of the young activists coincide with those who worked for the Student Nonviolent Coordinating Committee.

Meridian does not trace the history of the movement in an orderly, chronological fashion, but moves forward and backward in time, presenting a series of highly visual vignettes. Some of the novel's thirty-four pieces focus on public events: people singing, marching, being beaten, going to jail; the funerals of John Kennedy and Martin Luther King, Jr.; a political sermon in a black church. Others are highly personal: one girl having an abortion; another riding south in a car with her boyfriend; a man and a woman grieving over a dead child. Even the most private experiences in this novel have public, political implications, and the public events are presented not in terms of the large sweep of history but as the personal experience of individuals.

The absence of an interpretive narrative voice to make connections between the individual vignettes requires readers to fill in from memory the missing details, which in 1976 would have still been quite fresh. It had, after all, been only eight years since the funeral of Martin Luther King, Jr., was flashed across television screens around the world. The whole movement as it was experienced by television viewers was one of high points punctuated by sensational acts of violence. *Meridian* focuses on activists who did not have time to watch the evening news to see what was happening in Birmingham or Selma and who experienced the movement as a series of disconnected episodes interspersed with disabling personal conflicts, rather than as an interesting narrative smoothly moving from one stage to the next. Knowing that *Meridian* is about civil rights activists, readers might expect it to be about their contributions to the movement, but in fact this novel is more about what was lost than what was gained.

The form of *Meridian* is not that of a political or historical novel that sequentially traces social progress. In searching for a structure for *Meridian*, Walker explains, she "wanted to do something like a crazy quilt . . . something that works on the mind in different patterns."[3] The image of the quilt is spatial, and in *Meridian* the "scenes"—mainly from the civil rights movement and from the lives and family histories of the young activists—are arranged in a spatial rather than a temporal structure, despite the historical emphasis. Its thirty-four pieces are put together without apparent regard to chronology. Scenes from the mid-sixties are next to scenes from the mid-seventies; the events of the civil rights movement are juxtaposed with vignettes from the heroine's childhood and pieces of family history, folklore, myth, and current events. There are tales within tales—scenes within scenes—and it is sometimes difficult to see a relationship among the pieces. Yet there is a kind of order in this chaos. Walker explains that there is a significant difference between "a crazy quilt and a patchwork quilt," since the patchwork is literally just patched and the crazy quilt is planned.[4] *Meridian*, then, is a deliberately patterned work, made of seemingly different and not necessarily harmoniously arranged "scenes."

The impact of a crazy quilt depends on shapes, colors, and textures, but what usually gives such quilts significance and emotional validity is that each individual piece has a history: a blue patch might be from an old man's overalls, a red one from a party dress, and a pink one from a child's playsuit. The woman who owned the red dress may not be related to the man in blue overalls. Fragments might come from discarded clothes of the quilt maker's family members or the ragbags of friends. Kin and neighbors may recognize the origin of patches, recall their history, and so "read" the apparently "crazy" quilt that is meaningless to outsiders. The fragments of *Meridian* are also bits of history.

Some pieces are not stitched through narrative comment to the core narrative. The piece entitled "Gold," for example, relates the time when seven-year-old Meridian finds a large bar of gold. When her parents ignore her and refuse even to look at it, Meridian buries the gold and forgets about it. End of story. Readers might be tempted to interpret the story symbolically: the gold is Meridian's talent and intelligence, which her parents do not recognize and which she therefore buries and forgets, or the gold represents something of value that Meridian finds—the civil rights movement, perhaps—but which her parents refuse to affirm. But the novel does not offer an interpretation, and in fact it is never mentioned again. "Gold" is a discrete vignette, a brightly colored piece of fabric in the quilt, a part of Meridian's life story that is not echoed in any other part.

Meridian may be like a crazy quilt, but as a verbal structure it is more complex and intricate than a cloth quilt could ever be. Not only are there scenes within scenes—stories within stories—but there are pictures within the pictures that the stories make. The novel seems more like a firsthand documentary of the civil rights movement that presents raw history in a nonnarrative, spatial structure.

It may be useful then to reconstruct the primary or core narrative that is subtly embedded in the scenes: Meridian Hill, a bright teenaged black girl, whose young husband has abandoned her and their baby, watches television and learns about voter registration drives, a bombing in the local movement headquarters, and the deaths of children. Responding to these events, occurring "in the middle of April in 1960" (73), Meridian volunteers to work at the local movement house "typing, teaching illiterates to read and write, demonstrating against segregated facilities and keeping the movement house open when the other workers returned to school" (85).

Sometime in the summer—probably 1961—Meridian accepts a college scholarship from a white family that contributes to the movement by sending "a smart black girl to Saxon College in Atlanta" (86). After concentrating on her studies for a year, she joins the movement in Atlanta and falls in love with a young activist, the handsome Truman Held. Together they become acquainted with Lynne Rabinowitz and other white students from the North.[5] In the fall of 1963, "during the first televised Kennedy funeral" (33), she makes friends with Anne-Marion, a fellow activist. She canvasses voters, marches in the streets of Atlanta, sees old women beaten, and "men brandishing ax handles" chasing small children. Frequently jailed and beaten, "once into unconsciousness," Meridian begins to suffer as well the consequences of the stress, made worse by the turbulence of her relationship with Truman. After only one sexual encounter with him, she gets pregnant, has an abortion, and has her tubes tied—all in a matter of weeks. Further complicating her feelings is Truman's involvement with the white students, particularly Lynne Rabinowitz, whom he later marries. For Meridian, and for the reader of the crazy quilt, life in the civil rights movement seems "fragmented, surreal" (96).

Meridian's activities after she leaves college are not all precisely dated, but much of the chronology can be inferred. While sharing an apartment with Anne-Marion in Atlanta, she develops a debilitating illness. Around 1966, she is in New York with movement colleagues who demand that she take a vow to "kill for the Revolution" (27); refusing to do so, Meridian goes off on her own to work with poor blacks in small Southern communities. Truman and Lynne periodically visit her and seek her help with their various crises. In the mid-1970s, when Meridian is living in the small town of Chicokema, "near the Georgia coast" (144), Truman pays her a visit, and while he is there, she finally recovers her health.

The opening passage of the novel recounts Truman's arrival in Chicokema; the last tells of Meridian's departure. Left alone, Truman climbs shakily into her sleeping bag, grieving, thinking that "perhaps he would" resume her work there (220). All the past, Meridian's early life, her family history, the story of Truman and Lynne, and the civil rights movement itself, are placed between the accounts of these two events.

The novel is divided into three parts: "Meridian," "Truman Held," and "Ending." The first and third parts focus on Meridian, and the second on Truman and Lynne, whose experiences often intersect with Meridian's. After their days together in the movement in Atlanta, Truman and Lynne return to the North and secretly meet in Truman's parents' home. Horrified that she is involved with a black man, Lynne's parents disown her. Sometime later, Truman and Lynne move to Mississippi. It is not clear exactly when they marry, but some three years afterward, Truman, having tired of Lynne, periodically leaves her to visit Meridian. Paralyzed by fear of being thought a racist if she denies sex to black men, Lynne submits to "rape" by one of Truman's acquaintances and endures the sexual assaults of others. Shortly after Lynne becomes pregnant with Truman's child, she sends Truman "back to Meridian, at his insistence" (166), and returns to New York, where she lives on welfare and becomes fat and depressed. Truman sets up a studio in New York and becomes a successful painter. When a street hoodlum murders their daughter, Meridian comes to nurse Truman and Lynne through their grief. At the end of the novel, Truman tries to atone for what he has done to Lynne by telling her he wants to provide for her, like a "brother" (215). Interleaved with the saga of the Meridian-Truman-Lynne triangle are stories of the poor, the disenfranchised, the sick, and the uneducated, whose limited lives have hardly been affected by the social changes brought about by the movement and whose welfare first Meridian and finally Truman are working to ensure.

Though little of substance has changed for society's outcasts, much has altered—though not necessarily for the better—in the lives of the young activists. Truman, who once earned money working in an Atlanta country club serving white people, now drives his own Volvo and is making a sculpture for the Bicentennial. The former revolutionary Anne-Marion, now a well-known poet, writes about "the quality of light that fell across a lake she owned" (201). In contrasting scenes, we see Lynne, once slim and vivacious, now fat and depressed; Truman, once arrogant and confident, now tearful and uncertain; Meridian, once sick and helpless, now strong and purposeful.

In general, then, the core narrative is about a black man and the two women in his life—one black and one white—and how in the mid-1970s their lives have been transformed by their shared experiences in the civil rights movement. The first and third parts contain much more than the vivid scenes from the core narrative, however; there are highly visual, dramatized scenes from Meridian's childhood, legends of her family history, accounts of her adolescent sexual encounters, lore of Saxon College, and tales of the daily struggle of those working in voter registration drives.

The process of putting together again the humpty-dumpty that is *Meridian* is necessarily partial since readers are left in the end with many odd bits and pieces that do not contribute to a chronological rendering of events or connect with other fragments in any meaningful way. A sequentially structured, chronological narrative like Margaret Walker's *Jubilee* suggests that history is linear and progressive with effect following cause in an orderly fashion. A text structured from fragments that are not grounded in narrated history portrays experience as chaotic, fraught with conflict, and rarely smoothly progressive. Readers who are searching for pattern will inevitably reconstruct a narrative from the raw material, searching for sequence, for cause and effect, but *Meridian*'s fragmented form interferes in that process. Without a clear sense of historical cause and effect, *Meridian* lacks a mandate for the future, leaving readers to struggle with what comes next in the history of the struggle for racial justice. By declining the moral authority of a narrative overvoice or of a character whose choices are affirmed, Alice Walker forces her readers to make judgments that narrators sometimes make and that Shug Avery makes in *The Color Purple*.

When *Meridian* was published in 1976, this country, still reeling from Vietnam and Watergate, was making little noticeable progress in civil rights. The contribution that Meridian and others like her were making to the tedious work of the Voter Education Project, headed at that time by John Lewis, to whom the book is dedicated, was slowly yielding results, as increasing numbers of black officials were being elected to public office.[6] Ten years after the publication of *Meridian*, the then "unsung" John Lewis, one of the heroes of the movement, was elected to the United States Congress from Georgia through his own efforts and those of the unknown activists who continued the struggle after its days of glory had passed. The ongoing election of black public officials and the influence of the Reverend Jesse Jackson have in a sense validated Meridian's choice, which in 1976 might have seemed more problematic than it does today.

By using Meridian's name as the title of the novel, Walker invites readers to consider her as "prime," the meridian from which all else is measured. But Meridian's way is only one way, and significant portions of the novel concern alternative lives and views, some perhaps more suitable for the less saintly among us. All of the young activists have paid dearly for their involvement in the movement; none of the characters has smoothly paved roads ahead; all live with and expect to continue to endure conflict. None finds easy ways to negotiate the interaction between private and public aspects of life. For a brief time in the early days of the movement, Meridian experiences the feelings of solidarity and absolute commitment, untainted by contradictory claims: as she is being arrested and beaten, Meridian realizes that "they were at a time and place in History that forced the trivial to fall away—and they were absolutely together" (84). It is not long, however, before the drama and the glory and the unifying force of the movement are gone for all.

For Meridian and Lynne, joining the movement results in a rending with the past. Disowned by her parents and eventually rejected by the black community, Lynne ends as an outcast from both races. Meridian's mother is clear about where she stands on civil rights: "God separated the sheeps from the goats and the black folks from the white. . . . It never bothered *me* to sit in the back of the bus" (85). Fear of her mother's criticism does not deter Meridian from responding to the compelling mandates of the movement, and when the opportunity to leave her child and go to college arises, she accepts it in spite of her mother's profound disapproval and her own tortured nightmares of her baby "suffering unbearable deprivations because she was not there" (91).

For some time after she breaks with her mother and her child, Meridian feels that the past pulls "the present out of shape" (91), but eventually her break with the past is complete. In a scene toward the end, she attempts to "rouse her own heart to compassion for her son," whom she abandoned when she was seventeen years old, only to find that "her heart refuses to beat faster, to warm" for her lost child (213). Lacking an authoritative voice, the novel does not judge Meridian's action or her lack of feeling; rather, it dramatizes the power of public commitment to overwhelm the demands of private ties.

Meridian, Truman, and Lynne have all paid high prices for their roles in the movement. Not only have they endured the conflicts of the public and private, but they have lost what many people consider the focus of private life: children, parents, personal love. Rootless and homeless, they must find their way into the future without the stability provided by a sense of connection to a history in which cause and effect lead inevitably to a predictable future. Although it is clear that the two women made an irreparable and rending break with the past and with families that refuse to acknowledge the value of their lives, the novel is not so specific about the price Truman has paid. But it is clear that he is tortured and uncertain about how to live. Like the patches of a crazy quilt, the episodes in the lives of these young people have been torn from family and community, and each must create a pattern without the support of that context.

In order to engage in the intense political struggles of the movement, Meridian has to forget the events in her personal past that once kept her from the larger historical context of her life. She even finds it necessary to give her baby away. The historical context for Alice Walker, however, is not only that of traditional or political history but of what we might call cultural history. That history must not be forgotten. In an interview in which she discusses the structure and significance of *Meridian*, Walker speaks of her fears about "how much of the past, especially our past, gets forgotten."[7] This does not refer so much to events such as those that led from *Plessy v. Ferguson* to *Brown v. Board of Education*, but to the substance of folk life: its legends and myths, its arts and symbols, its morals, customs, and habits—all of which may or may not change in response to those public events often equated with history.

In an essay written in 1970, Walker wrote of the importance of ordinary black Americans having a sense, not just of public history, but of continuity and context, of learning to "see themselves and their parents and grandparents as part of a living, working, creating movement in Time and Place"—that is, temporally or historically, as well as spatially.[8] Thus, when she writes of Saxon College's "long, placid, impeccable history" (48), Walker is referring to the kind of homogeneity of experience that has characterized the institution; she does not mean its impact on external events or the nature of society at large.

In a rebellion against the college's authority figures, who have reprimanded them for their activism, the students respond by destroying the "Sojourner," a giant magnolia tree, long a symbol of cultural, mythic history. Since legend says the tree grew from the tongue of a slave woman, they literally sever a living link with the past. Ironically, taking a public stand results in the students breaking with the cultural history that strengthens them and makes activism possible. By joining the civil rights movement, many young people—black and white—placed a permanent barrier between themselves and their heritage, cutting themselves off from their cultural (or, in Walker's terms, historical) roots—the morals, customs, and habits of their families and communities, as well as their own personal or sequential histories.

The cultural history that Walker treasures is represented in visual images that inspire the young activists. It is through pictures, of course, that Meridian first gets involved in the movement: watching television, she is inspired by the pictures of the struggle; at the same time, a white family in Connecticut sees the same pictures of "courageous blacks . . . marching and getting their heads whipped nightly on TV" and responds by sending Meridian to college (86). Throughout *Meridian*, pictures give their mute testimony. Among the restraints that prevent her from endorsing violence are memories, mental pictures of "old black men in the South who, caught by surprise in the eye of a camera, never shifted their position but looked directly back; by the sight of young girls singing in a country choir" (27–28).

In Truman's mother's house, there are paintings by Romare Bearden, Charles White, and Jacob Lawrence, all black artists. As an artist, Truman is, of course, always making pictures. When he and Lynne live in Mississippi, for example, he paints a mural of the civil rights struggle. He also creates a series of paintings of Meridian's dark brown face and other black women depicted "as magnificent giants, breeding forth the warriors of the new universe" (168). When he and Lynne go south together, among his few possessions are two cameras (154); at one point, he frames a picture of Lynne—sitting on the porch of a shack, surrounded by black children—but then stops suddenly and takes instead "a picture of the broken roofing and rusted tin on wood that makes up one wall of a shabby nearby house" (129). Anne-Marion sends Meridian a photograph of the stump of the Sojourner, with "a tiny branch . . . growing out of one side" (217). A grieving father stands before the congregation of a black church beside a photograph of his son, a slain martyr in the civil rights movement (194). Lynne is pictured "nestled in a big chair . . . under a quilt called The Turkey Walk" (130), and when she grieves over the loss of her daughter, she finds comfort by spreading the same quilt over her knees.

Having concluded that the church is "a reactionary power" (199), Meridian is surprised to see a stained glass window in a church that depicts "a tall, broad-shouldered black man" dressed in brilliant red and blue, with a guitar attached to a golden strap in one hand and a "shiny object the end of which was dripping with blood" in the other (198). When Meridian asks the rather conventional woman sitting next to her what it is, she casually responds, "Oh, *that.* One of our young artists did that. It's called 'B. B., With Sword'" (199).

Pictures were, of course, the impetus for many episodes of the civil rights movement. Pictures of Emmett Till's mutilated body in *Jet* magazine in 1955 moved blacks across the nation to send money to civil rights organizations, demanding that something be done; those of demonstrators attacked by dogs in the streets of Birmingham in 1963 outraged and activated still others; that of John Lewis being beaten on the Edmund Pettus Bridge sent thousands to Selma.

Not all the novel's strong visual images are found in the description of paintings, photographs, stained glass windows, or quilts. Equally powerful are the verbal pictures created by highly visual vignettes that are in a sense framed by their isolation from the progressive narrative. The ongoing suffering of impoverished black families in the seventies is evoked through vignettes that lead readers to "see" Meridian placing beside the mayor's gavel in a public meeting the decomposing body of a small child, drowned because of public negligence; the woman dying without proper medical care lying beneath a "faded chenille bedspread," her son who "don't have shoes" cuddled beside her (204–05); the round and clean baby strangled to death "with a piece of curtain ruffle" by its thirteen-year-old child mother (212). Lacking narrative buildup or consequences or the welcoming threshold that tells readers what to think, these vignettes illustrate the continuing suffering that occurs in a society indifferent to its poor. They depict the unfinished business of the movement in sparse verbal pictures unencumbered with narrative interpretation. Readers must reach their own conclusions about what it means that former civil rights workers are still struggling in the mid-seventies to register voters and are limited to such small victories as having a drainage ditch filled or adding another name to the list of potential voters.

There are a few passages, however, that make the relation of art to social action problematic. In the opening piece set in the mid-1970s, Meridian asks Truman if he is a revolutionary. Truman answers: "Only if all artists are" (24). Lynne feels guilty for thinking that "the black people of the South were Art" (130). Art functions in this novel, then, only to confront its audience with the unmediated, and therefore realistic, scenes, however fragmentary and incomplete, of specific social conditions. By portraying—picturing—injustice and other social ills, or passing on cultural history, art can engender social action, though it does not exhort, direct, or define what action is to be taken.

The novel's unresolved political issues demand as much of readers as does inferring the social function of art. In the beginning and ending scenes of the novel, which take place in the mid-seventies, Meridian is still obsessed with the question of whether she can "kill for the Revolution" (27), an issue that was first raised in the movement and then directed specifically at Meridian in the sixties by her friend Anne-Marion. Musing on whether her friends, "a group of students, of intellectuals, converted to a belief in violence," could actually face "the enemy, guns drawn," Meridian concludes, "Perhaps. Perhaps not" (28). Her ambivalence persists; she alternately feels that "indeed she *would* kill" and that such resolve is prompted by "false urgings . . . in periods of grief and rage" (200).

Conflicts about whether violence is an appropriate response to racism divided members of the various civil rights organizations and "perhaps" even brought an end to effective coalitions among those groups. By the time Meridian meets Anne-Marion in 1963, the role of violence had already become a major subject of debate among civil rights leaders, including John Lewis, to whom the novel is dedicated. As the chairman of the SNCC, Lewis was pressured to tame the potentially inflammatory rhetoric of the speech he had prepared to give at the great March on Washington in August of that year. By 1966, however, the issue of nonviolence versus armed self-defense had begun to divide the leadership of SNCC, and it was Lewis's insistence on the importance of maintaining a stance of nonviolence that led to his resignation and to the rise of black power and its dominant advocates in SNCC, first Stokely Carmichael and then Rap Brown. At about the time that Lewis took his stand and resigned from SNCC, Meridian was heading south, having taken the same stand with her colleagues in the movement.

In the final scene of the novel, her health restored and her commitment made to "return to the world cleansed of sickness" (219), Meridian feels that she is able to bear "the conflict in her own soul" (220), though she still alternates between the fear that she will "not belong to the future" and the tentative consolation that "perhaps" it will be her role to "walk behind the real revolutionaries . . . who know they must spill blood in order to help the poor and the black . . . and sing from memory songs they will need once more to hear" (201). Perhaps Meridian expects to play the role of poet and storyteller, following the choice that Walker herself has made in singing the songs she feels her people need to hear, providing in literature those models that are in fact lacking in so many people's lives.

Spatial in conception, fragmentary in structure, visual in impact, *Me-*

ridian evokes the civil rights movement, while refusing to reach simplistic conclusions or to forge an orderly and therefore falsifying causal narrative. Its tentative, inconclusive qualities underscore the inaccessibility of the period and suggest that the lessons of history cannot be systematized and that their influence is easy to exaggerate. *Meridian*'s unremitting ambivalence is reinforced by the frequently repeated word *perhaps.* Perhaps Meridian will find a way to play a significant role in the future; perhaps Truman will renounce his self-indulgent ways and find a way to relate his art to society; perhaps Truman's offer to take care of Lynne like a brother will provide her with an opportunity to recuperate from what seem to be irreparable scars and to find a way to begin her life anew. Perhaps.

Ending with Truman's recognition that Meridian's "sentence of bearing the conflict in her own soul . . . must now be borne in terror by all the rest of them" (220), the novel suggests that liberty and justice for all is at best a fragile prize to be won—and paid for—again and again. In the context of the Bicentennial celebration, *Meridian* is a powerful reminder that, for those who are committed to a just society, the struggle will be slow, tedious, conflict ridden, and lifelong.

The Salt Eaters

In its content as well as its temporal and geographical setting, Toni Cade Bambara's *The Salt Eaters* is more like *Meridian* than any other novel considered here. The primary narratives of both novels take place very near the Bicentennial year: In Walker's novel, Truman Held is making a sculpture of Crispus Attucks—the ex-slave who was a hero of the American Revolution—for the 1976 celebration. Bambara's Dr. Meadows has just turned down the part of Attucks in a Bicentennial pageant after learning that because of his light skin he will have to play the role in blackface. Set primarily in Georgia, each novel has a female protagonist suffering from debilitating illness that is at least in part caused by the stress of the civil rights movement. At the end of both novels, the protagonist, restored to health and well-being, is facing a difficult future. On the final page of *Meridian,* Truman feels that the conflict Meridian has endured "and lived through—must now be borne in terror by all the rest of them" (220); the narrator of *The Salt Eaters* remarks at the end that Velma, imagining she is having a hard time, "didn't know the half of it. Of what awaited her in years to come" (278).

Like *Meridian*, *The Salt Eaters* focuses on the public and private lives of civil rights activists struggling to find their way in the post-movement period, while still nursing wounds sustained "in the days of the marching" (33). While Shange's *Betsey Brown* suggests that, with enough love and good humor, the conflicts between public and private life can be resolved, and *Meridian* stresses the inevitability of such conflict and the necessity of enduring it, *The Salt Eaters* assumes that a full life includes the healthy integration of private and public commitments and that, when one is seriously disturbed, the other will suffer. Whereas Meridian recovers her health, apparently through her own efforts, Velma becomes well again through the massive intervention of the entire community. For Bambara, the health and well-being of individuals is the responsibility of a healthy community.

Coming to the end of Toni Cade Bambara's complex and provocative novel, a reader might appropriately declare, in the words of one of the characters, that it "whelms me over" (162).[9] More complex than that of her own editor Toni Morrison, more probing than that of Alice Walker, Bambara's work is best summed up with words she uses to apply to one of her characters: "so tough, so compassionate, so brave." "Tough" is the word that best describes the prose and elusive structure of *The Salt Eaters:* tough to read, tough to digest, and even tougher to assess. Like Morrison, Bambara sets her characters squarely in history and dares to combine realism and magic; like Walker, she treats the civil rights movement without sentimentality. Unlike Morrison, however, she creates a host of characters who live deliberately and consciously in the public arena with an awareness of history; and she shuns the grotesque, the morbid, and the perverse.[10]

Before she settled on *The Salt Eaters,* Bambara had other working titles, including "In the Last Quarter," referring to the last quarter of the last century of the millennium. Though she abandoned that title, Bambara wove references to the last quarter throughout the novel, reminding readers of the possibility of long-predicted cataclysmic events and suggesting that the time of the novel is the readers' time as well. The title *The Salt Eaters,* however, focuses the readers' attention on those who are seeking health rather than on those who are doomed. Salt eaters are people who still practice the old folk medicine, including eating salt as a cure for snakebite. The cure the novel seeks, however, is for the disorders caused by poisons that afflict the characters' minds, and by extension the society they create. Bambara has written elsewhere about her decision to exclude from her work "rage, bitterness, revenge," which she calls "poisons," and about how if people are to "struggle, to develop," they need "to master ways to neutralize poisons."[11] As a remedy for the poison in us all, *The Salt Eaters* sometimes goes down like brine.

The main action of the novel—the healing of Velma Henry—takes place in the imaginary town of Claybourne, Georgia. Velma has attempted suicide. Throughout the day and night in the spring of 1977 that make up the primary time of the novel, she sits on a stool in the Southwest Community Infirmary attended by Minnie Ransom, an old "fabled healer of the district" (3) and surrounded by a dozen or so observers who have come to watch Minnie work. What is wrong with Velma, the narrative reveals, is also wrong with the community. Referring to the disease that plagues the community, one character observes that "Claybourne [is] getting to resemble the back wards of the asylum more everyday" (216). Those who participate in healing Velma heal themselves in the process.

In *The Salt Eaters* the health, well-being, and welfare of individuals are determined by their relation to family and friends, as well as to their work, the community, and the personal and communal past. To be well requires getting outside the self, understanding the interrelatedness of private experience and public life, and acting accordingly. To get outside the self, however, involves confronting simultaneously the consequences of the past and the requirements of the future. Velma's illness is at least in part due to her attempt to withdraw from her public roles into the "self," a world so personal that all communal, public existence ceases to be: "She tried to withdraw. . . . Withdraw the self to a safe place where husband, lover, teacher, workers, no one could follow, probe. Withdraw her self and prop up a borderguard to negotiate with would-be intruders" (5). It is Velma's urge to pull back from her social responsibilities—as mother, wife, worker, activist—that must be cured in order to restore her to society and to health.

Bambara captures the enormously complex relationship of the individual to the community, the private to the public sphere, by creating a multifaceted, many-layered narrative form. Much of the narrative consists of Velma's memories of the past and those of the characters observing Minnie's efforts to heal her. The point of view shifts from that of the objective narrator to that of various characters. There are flashbacks and what Bambara calls "flashforwards," as well as flashbacks within flashbacks.[12] This flexible form allows Bambara to place her characters in a context that is temporal, geographic, personal, and public.

Epic in scope, *The Salt Eaters*, divided into twelve chapters, includes many characters struggling with conflicts of public and private life; most are identified by their vocation or social roles: artist, politician, trade unionist, bus driver. By the end of this relatively short novel—less than three hundred pages long—there are more than seventy-five named characters: journalists, engineers, musicians, doctors, activists, teachers, as well as a masseur, a dance instructor, a trade union leader, and a jewelry maker. Indeed, the novel typically reveals the characters' public roles before their private ones. For example, we learn that Sophie Heywood is and has been "the chapter president of the Women's Auxiliary of the Sleeping Car Porters for two decades running" (12) before we learn about her husband or son; Marcus Hampden is identified as "a member of the Coalition of Black Trade Unionists" and later as Palma's lover.

Many of the characters are little more than a name and a vocation; others come with details that define their mental landscapes. For example, Velma is a computer analyst who may be sabotaging the data of the Transchemical Corporation that is polluting the neighborhood; her husband is sleeping around; she feels alienated from her son. Fred Holt, the bus driver, has marital problems, is afraid of losing his job, and is grieving for a friend who was senselessly murdered. Dr. Julius Meadows, the young middle-class physician, alone and aloof, feels alienated from the black community, while Campbell, newspaper reporter and part-time waiter, imagines that someday he will get a break and write "the big story" (209). Minnie Ransom, who has the first and almost the last words, is profoundly troubled about her role as a healer in a disintegrating community. Yet she continues to seek ways "to pull it all together and claim the new age" (46).

Near the infirmary is the Academy of Seven Arts, a kind of community center, where Velma and her husband, Obie, direct the various projects that have evolved from the civil rights movement.[13] Alternating with the scenes in the infirmary and the academy are episodes in Fred Holt's bus, on the street, and in a cafe. What brings the characters together in these three settings is a carnival, a small town Mardi Gras that has brought crowds of people into the streets, the park, and the town's one stylish cafe. Fred Holt is driving a bus whose passengers include activists planning to join others in the cafe and in the carnival celebration that is about to begin. Dr. Meadows wanders the street among "the children out of school; adults out of work; and workers from the chemical plant's second shift, smudged and smelly and in search of a quick bite to eat" (123). Campbell overhears and periodically joins in the endless table talk. Although there are occasional encounters between only two people, almost all the scenes include larger groups. Chapters 1 and 2 introduce the characters in the infirmary and through a series of flashbacks connect the present with the personal and public past. Chapter 3 shifts to a bus, en route to Claybourne, and focuses on passengers struggling with their current personal and public dilemmas.

But for Fred Holt, the driver, the past relentlessly intrudes on the present. Among the passengers are seven young women, antinuclear activists, returning from a demonstration. While eavesdropping on the women's talk about Bakke, Carter, the KKK, and International Women's Day, Fred Holt begins to think of his friend Porter, a "race man" (82) doomed by his exposure to atomic blasts in 1955 but murdered by a crazy woman with knitting needles. Fred's memories of Porter alternate with thoughts of his childhood, his marriages, and the world around him. When he sees "old men in tatters," he thinks "it could be the Depression again" and fears unemployment, inflation, and a return to poverty (70). Fred's musings alternate with those of his passengers, who drift into thoughts about their own personal and public pasts.

The shifting focus of chapters 4 through 7 fills in details about other characters, including Dr. Meadows, who has come to observe the healing of Velma. Wandering through the streets of Claybourne, he overhears politically charged talk and a street speaker calling for "transformation . . . [and] new alliances" (126). Suddenly he is assaulted by memories of the civil rights movement. Recalling "Greensboro, Montgomery, Port Gibson, Little Rock, Hattiesburg" (177), he acknowledges how limited his contribution to the movement had been. While others were marching and risking their lives, he had written checks. Encounters with an old man begging for his "suppah" (176), the "half-men" (182), and "the welfare mamas" (183) leave him significantly shaken and alienated from the black community that is his birthright. Frightened by a snarling dog and threatened by street thugs who call him a "honky" (185), Meadows realizes that he has paid dearly for his place in middle-class society, his expensive clothes, and Omega watch. The spring festival that had been planned "as a holding action" (92) against the community's rapid disintegration may have brought people together into the streets and public meeting places, but in this and other scenes it serves mainly to highlight the distances between individuals and the need for some new coalition.

Subsequent chapters include scenes with Campbell, the waiter/journalist presiding over the multi-ethnic crowd of engineers, writers, activists, and artists at the local trendy cafe. For the activists, the civil rights and the antiwar movements have been replaced by the women's movement and the antinuclear movement, yet the days of the marching seem present in the thoughts of those who have moved on to other causes. Much of the talk is argumentative and concerns controversial public issues: the disintegration of the movement, the nuclear threat, and women's liberation. Typical is the argument between two women, both civil rights workers. Irritable and depressed, Ruby, trying to explain to Jan what is wrong, begins with personal problems—her husband is on the road. Then she moves to what is wrong in the outside world: "Malcolm gone, King gone, Fannie Lou gone, Angela quiet, the movement splintered, enclaves unconnected" (193). Puzzled about "who could effectively pull together the folks" (193), Ruby is close to despair. When Jan urges her to come join the festival in the park—the festival that was to be a staying action against disintegration—Ruby declines: "I'm sick of all this pagan spring celebration shit . . . leaflets and T-shirts and moufy causes and nothing changing" (201). Ruby, like the community of social activists, is sick and tired—"tired of worrying about Velma," the invasive past, and "the wet tents and bloody feet" (216). The fragmentary, disconnected, antagonistic conversation of the diners is symptomatic of a community of individuals unable to move beyond their narrow, personal perceptions in the wake of their participation in the civil rights movement.

In chapter 11, the conversation resumes between Velma's friends, Jan and Ruby, who are waiting for her and do not know about the suicide attempt. Arguing about what might be wrong with Velma's life, Jan points out that Velma and her husband have "set things up" so they could never "opt for a purely personal solution" (240). For Jan, a "personal" life is one lived in "confined space, everything under your sure control" (240), and she argues that it is difficult "to maintain the right balance there, the personal and the public" (241). Jan seems to be saying that activists like Velma who live in open public space lack the private ballast needed to create whole, balanced lives. But Jan's voice is only one of many in *Salt Eaters*, and her conclusion serves merely to feed the multi-voiced discourse that is its center. Toward the end of the novel, Ruby complains about "this doomsday mushroom-cloud end-of-the-planet numbah" and urges her friend Jan to go back to the subject of race. Jan argues, however, that the problems are related, that the country's underclasses have suffered the most from the consequences of the arms race (242).

Meanwhile, during all these events, Velma sits on the stool in the infirmary recalling the private and personal worlds that overlap in her memory: the day she was baptized, an emotional encounter with her husband,

Obie, their courtship, and a recent political meeting. She remembers the movement as a time when the women were doing the dirty work while the men were "renting limousines" and strutting around in "three-piece suits" (37). And she still has tortured memories of the time she left hungry and sick protesters in their tents in the mud to search for help, only to find Obie in a luxury hotel with another woman at a time when he had told her he was going to Washington to meet with King. Images of the past invade the present for Velma, who recalls a painting of Martin Luther King, Jr., in a meeting room where the talk was about the Carter administration, Social Security, and South Africa; and a speaker at a rally who "looked a bit like King, had a delivery similar to Malcolm's, dressed like Stokely, had glasses like Rap" (35).

Gradually Velma relinquishes control and experiences the pain of memories that are as much public as private, and her response to the pain, a kind of guttural "growling," is the first step she takes toward health. Affirming the process, Minnie Ransom tells her: "Growl all you want, sweetheart. . . . Growl on. You gonna be all right . . . after while" (41). Midway through the novel, the omniscient narrator announces that Velma will not only recover but that her recovery will serve as an example for others: "in time Velma would find her way back to the roots of life. And in doing so, be a model. For she'd found a home amongst the community workers who called themselves 'political.' And she'd found a home amongst the workers who called themselves 'psychically adept.' But somehow she'd fallen into the chasm that divided the two camps" (147). In her struggle to bring Velma back to "wholeness," Minnie insists that Velma must want to be well, that wellness involves responsibility, and that she must decide what she will do when she is well again (220). Recovery in terms of *The Salt Eaters* is finally taking responsibility for personal and social health that makes it possible to climb out of the chasm and build a bridge between the public, political and the private, psychological worlds.

Trying to come to terms with his wife's attempted suicide, Obie is at the same time struggling to understand the fragmentation of the movement as it is reflected in the Academy of Seven Arts. Concerned about dissension within the ranks, Obie muses about the collapse of the community of social activists that make up his and Velma's world: "the factions . . . a replay of all the ideological splits . . . threatening to tear the Academy apart" (90). Alone with his thoughts, Obie searches his memory for a vision of wholeness and remembers Sophie Heywood's admonition: " 'There is a world to be redeemed,' she warned. 'And it'll take the cooperation of all righteous folks'" (92). Wholeness, cooperation, coalition, Obie feels and the novel shows, must come from within and without: "And so he stretched and breathed deeply, trying to pay attention to what he saw and heard and felt around him and inside" (98). The cure for political, social, and personal illness to which the novel comes is finally holistic: what is good for society is good for individuals, and vice versa. Those who live too exclusively in the public sphere will finally pay for that choice in their personal lives, and those who attempt to live in private domains, like Jan, will pay with the quality of their society and their limited private lives. To be effective, a healing process must encompass the psychological and the spiritual, the social and the political.

Minnie explains that to get well Velma must give up "the pain, the hurt, the anger" in order to make room for good feelings; she must give up stomping around in the "mud puddle" of the past (16). Velma begins to recall a recent time when Obie also accused her of hanging "on to old pains" (22) and begged her to "push all the past aside" and to "create a vacuum for good things to rush in" (25). To heal Velma, Minnie Ransom counters the disabling "flashbacks" of the past with "flashforwards." By balancing its tales of the past with visions of the future, the narrative holds the reader and the characters in the tension that is the present. Nothing in the narrative suggests that the characters might completely escape the past that invades and shapes the present. What happens in the novel and what Bambara presumably prescribes as a cure for the social sickness that plagues the novel's community and the reader's world is the creation of a balancing act between the memories of the past and visions of a future. To counter the power of the past, those visions must be fraught with urgency.

The Salt Eaters does not end with despair. Velma does get well; the community is revitalized. But how? The final chapters are increasingly characterized by what Frank Kermode calls "narrative anarchy."[14] Something happens that changes everything. But what?

As some of the characters pass the time in the cafe, others celebrate in the street, and still others watch the final stages of Minnie Ransom's healing, all activity is interrupted by a bright light, loud noise—perceived first as thunder—followed by a drenching rain. From the first flash of light, characters react to the event that "sounded like a rumbling of the earth, like a procession, like marching" (223) as if the world were about to end. One character recalls Hiroshima; another stares "at the ground as if waiting for it to crack wide open" (284). Campbell, looking back on that moment, remembers, "A grumbling, growling boiling up as if from the core of the earthworks drew a groan from the crowd huddled together under the awning, in the doorway, as if to absorb the shock of it, of whatever cataclysmic event it might turn out to be, for it couldn't be simply a storm with such frightening thunder as was cracking the air as if the very world were splitting apart" (245).

The moment, the beginning of a new age, readers are repeatedly reminded, "would be fixed more indelibly on the brain and have more lasting potency than circumstances remembered of that November day in '63" (246). This unspecified cataclysmic event, associated narratively with the assassination of John Kennedy, affects all, reinforcing the idea that, in the nuclear age, no one can remain entirely in the personal sphere. For Dr. Meadows, the event, whether cosmic, cataclysmic, or meteorological, is the apocalyptic beginning of a new life inaugurated with the "vow to give the Hippocratic oath some political meaning in his life" (281).

Just as Morrison leaves her readers to grapple with what will happen to her characters at the end of *Song of Solomon* and *Tar Baby*, Bambara tosses the what-has-happened ball into the reader's court. There are a number of realistic possibilities: an explosion at the Transchemical plant where Velma works, an accident at the nearby Savannah River plutonium production facility, or even a meltdown at a nuclear power plant. Or in her healing of Velma, Minnie Ransom might have harnessed cosmic forces capable of healing the entire community. Or it may be the Apocalypse, the triumph of good over evil.

Early in the novel, Minnie introduces the threat of Armageddon in a conversation with Old Wife, her spiritual guide from another realm. Insisting that there is an urgency, not just for her patient Velma Henry but for all civilization, Minnie explains: "Old Wife, we gonna have to get a mighty large group trained to pull us through the times ahead. Them four horses galloping already, the seven trumpets blasting. And looks like we clean forgot what we come to do, what we been learning through all them trials and tribulations to do and it's now" (46).

Though Minnie wonders how they will "pull it all together and claim the new age," she seems confident that they will. Other characters also have a basic confidence in their personal and communal survival. Even in the face of his wife's despair, Obie senses "a plan of growth for himself, for him and Velma, for the Academy, for the national community, for the planet" (98). The nature of the cataclysmic event at the end of the novel remains ambiguous, and readers are left to wonder what exactly has happened. But there is no ambiguity about there being a future on the other side of the event, as a number of characters look back to that day as the moment that a new "beginning was ushered in" (281). In a flash forward to the future—1984 to be exact—we see Dr. Meadows, transformed into a committed activist, talking with Velma's godmother, Sophie Heywood, who is explaining to him that "the second coming and Armageddon" should have been "translated 'presence of Christ' and 'New age'" (282). The novel ends, however, not in the future but back firmly in the present, as Velma Henry "rising on steady legs, throws off the shawl that drops down on the stool a burst cocoon" (295).

Like all apocalyptic fiction, the vision of Armageddon forces readers to consider what might be done to deter catastrophe. In *The Salt Eaters* Bambara counters the vision of Armageddon with a society somehow miraculously saved, clearing the space for readers to consider that healing of individuals and of society occurs not because known solutions are applied to diagnosed disorders but because enough people—a community—struggle to neutralize poisons, to balance the demands of private and public life, and "to be alert to the demands of time" (10). Though they bumble and fall, the characters of this novel, the salt eaters, may very well be the "mighty large group trained to pull us through the times ahead" (46).

Composed and sent to press during the Carter years, *The Salt Eaters* was created by Bambara as an antidote to "last quarter" post-Watergate despair. It is a Carter-era book—he is mentioned on page twenty-seven—speaking into a time that, compared to the Nixon years, was better, even benign. Yet there was no movement, no plan, and little activity on a large scale, and increasing emphasis on nuclear deterrence and anti-Soviet rhetoric intensified fears of a nuclear disaster. Still, renewed activity and communality seemed possible in this friendly if ineffectual atmosphere. To harness that spirit again would take something big—the healing of individuals in the context of community, the threat of an Apocalyptic storm. Perhaps the public spirit could be revived. After all, it had toppled Nixon and ended the Vietnam war. Perhaps it could do more. So it seemed in the late 1970s before the election of 1980.

Tar Baby

Toni Morrison's *Tar Baby*, published in 1981, the year after *The Salt Eaters*, focuses on individuals who have cut their ties with community, family, and the past. Fugitive, exile, expatriate, castaway—these are the roles the characters assume. When they come together on a remote Caribbean island, none takes any responsibility for the welfare of even a small personal community or attempts to be a part of the greater society. Torn by the inner conflicts of their own psyches as Bambara's salt eaters are immobilized by their conflicting public and private commitments, Morrison's characters pay the price of living self-absorbed lives, isolated from roots and cultural traditions.

The novel opens as an unnamed black man, later identified as a fugitive and a Vietnam veteran called "Son," steps off a boat into the warm Caribbean Sea and makes his way to "L'Arbe de la Croix," an elegant house inhabited by the seventy-year-old Valerian Street, a white man; his pushing-fifty, beautiful, pale, red-headed wife, Margaret; their eternally absent only son, Michael; their old family-retainer black servants, Sydney and Ondine; and their servants' light-skinned niece Jadine, a successful model in Paris. Also serving the household are island blacks—Gideon, called "Yardman"; Thérèse, the laundress; and Alma Esté, who keeps Thérèse company while she works.

The characters in *Tar Baby*, like those in *Song of Solomon*, seem singularly unaware of the outside world, but unlike *Solomon*, *Tar Baby* includes very few references to public figures or events. The infrequent mail to the island does not seem to include newspapers, and Valerian mainly reads seed catalogs. Although Sydney and Ondine have copies of the *Philadelphia Tribune* in their apartment, none of the characters ever refers to international events or news from the states, and the only reference to a public figure is a "Vote for Dick Gregory" sticker dating from 1968. Social activism enters the novel only in stories about Michael, who spends his time demonstrating for Indian rights, environmental protection, and other causes. Jadine remembers that Michael made her feel guilty for not working for the civil rights movement in the 1960s. No one else thinks beyond immediate self-interest. All make decisions that are based on personal whim and desires. And this occurs, appropriately enough, during the months that Americans were preparing to put Ronald Reagan in the White House. *Tar Baby*, like *Song of Solomon*, is about living entirely in the private sphere; unlike *Solomon*, however, characters largely succeed in avoiding all consciousness of public life—on the island, in New York, and in the small all-black community called Eloe where Son grew up.

Retired from running the family candy factory in Philadelphia to his luxurious island home on Isle des Chevaliers, Valerian drinks wine, listens to classical music, and tends his own garden—that is, the plants in his greenhouse. Sydney and Ondine wait on him and run the house; Margaret comes for extended visits; Michael periodically promises to come but never does; and Jadine drops in from Paris and keeps Margaret company while desultorily doing paper work for Valerian, who has paid for her education and continues to provide her with money. Son—uneducated, brash, and on the run from the law—prowls about the grounds at night, steals food and, like Brer Rabbit in the Tar Baby folktale, helps himself to bottles of water.[15] Eventually, he extends his roaming to the house, where Margaret finds him hiding in her closet.

Son's appearance breaks down the precarious equilibrium of the household, turning it topsy-turvy. Valerian, presumably to provoke his terrified wife, invites him to stay in the guest room, and the members of this unusual household debate what they should do with this strange unkempt man. But their questions are soon rendered irrelevant as the next morning Son bathes, dresses in attractive clothes that he borrows from Valerian, begins a serious flirtation with Jadine, and generally makes himself at home. When Christmas day arrives, Son takes the absent Michael's place at the table, and Sydney and Ondine step out of their roles as servants and join the others at the table as guests. Ondine throws the household into another uproar by unexpectedly revealing the long-kept secret that Michael avoids his parents because Margaret abused him when he was a small child. Strongly attracted to each other, Jadine and Son take advantage of the chaos, spend the night together, and run off to New York, where they feel like "the last lovers in New York City—the first in the world" (229).

During this idyllic period, differences of class, education, and even race melt away in the heat of their loving, as Son makes Jadine feel "unorphaned" (229). For several months Jadine and Son are absorbed with their passion and with the excitement of getting to know each other. She "pours out her heart," telling him "secret things"; Son, who cannot talk "coherently" about Vietnam, tells her "what she wished to hear about the war" (224). They tell "each other everything" (225) and, in the anonymity of Manhattan, seem to embrace each other's differentness. When they visit Son's hometown of Eloe, Florida, however, disparities of race, class, and culture begin to cause trouble.

Eventually Jadine and Son can no longer gloss over contrary views of the black and white worlds. Jadine has white lovers and has spent much of her time in Paris living and working in white society, alienated from other blacks. A white lover, who wants to marry her, sends her a coat made of the skins of "ninety baby seals" (87). Son wants nothing to do with whites and seems to feel most at home with the people from the community where he grew up. In the small room in Aunt Rosa's house in Eloe where Jadine and Son spend a night together, a vision of chiding, critical "night women" invades Jadine's consciousness: all of them—including Son's dead wife, her own dead mother, and Ondine—stare at her and condemn her. Within five days among his provincial, poor, and uneducated friends and relatives, Jadine feels "orphaned again" (260) and returns to New York alone. When Son rejoins her, the harmony of their early love is broken, but their frequent fights seem to grow more from the clash of cultures than from opposing personal needs.

Race and class prejudice also affect other characters: part of what divides Margaret from Valerian is her inferior social status. Born to working-class parents, Margaret is more comfortable with her black servants than with her husband or his sophisticated friends. Sydney and Ondine look down on blacks that they think of as their social inferiors. Ondine insists that Son is not even "a Negro—meaning one of them"—rather he is a "nigger," who is "not going nowhere" (102). Sydney, himself "one of those industrious Philadelphia Negroes—the proudest people in the race" (61), refers to Son as a "stinking ignorant swamp nigger" (100). Yet Sydney lacks confidence that Valerian will take care of him in his old age and warns Son that "white folks play with Negroes" (162). Jadine, after her first encounter with Son, chats carelessly with Margaret about him, calling him a "nigger," though she bristles when Margaret extends the insult by saying he looks like a "gorilla" (129). Like her aunt and uncle, Jadine has learned how to get what she wants from white people—to say what they want to hear and in various ways to set the terms of their interaction.

Jadine, with skin the color of "raw silk" (272), and Son, "black as coal" (220), represent the opposite ends of the African-American spectrum on issues of color. She, having moved in an integrated society of "Blacks and whites in profusion," knows how to behave to be accepted by ambitious cosmopolites of both races: "She needed only to be stunning, and to convince them she was not as smart as they were" (126). Son, on the other hand, looks at Yardman's naked back bent in work, feels the pains and deprivations of his race, and thinks of privileged white children playing tennis under a sun "whose sole purpose was to light their way, golden their hair and reflect the perfection of their Easter white shorts" (139). Scornful of the white world, Son insists that "white folks and black folks should not sit down and eat together" (210); yet he is powerfully attracted to this woman the island blacks think of as "yalla" (155) and who lives mostly in a privileged white world. These differences are not resolved, as they continue to fight over her allegiance to the cosmopolitan, urban, and integrated but largely white world; his attraction to the rural culture of his birth; and their mutual inability to bridge the chasm that divides them.

Jadine and Son fight about race from the moment he invades her bedroom. Son accuses her of being like "little white girls" who always think "somebody's trying to rape" them, and when she insists that she is not white, he urges her to "settle down and stop acting like it" (121). But Jadine longs to escape race, culture, and history, to get out of her skin to be "not American—not black—just me" (48), and when Son refers to Valerian as "that white man" (263), she counters that he is "a person, not a white man" (263). For Son, Jadine's desire to be assimilated is a betrayal of her racial identity. He insists that accepting money from Valerian is equivalent to collaboration with the enemy and that if she continues to adopt the trappings of the white culture she might as well marry her former white lover and have his babies, doing what black women have always done: "Feed, love and care for white people's children" (269). For Son there is no compromise: "People don't mix races, they abandon them or pick them" (270).

Yet for all her arguments about the irrelevance of race, Jadine is uncomfortable about her choices. The nightmare of the chiding "night women" that drives her from Eloe has the same emotional content as an incident in Paris. While shopping for groceries, a tall woman with "skin like tar" in a canary yellow dress with eyes "too beautiful for lashes" stuns the other shoppers with her "transcendent beauty"; then she floats out of the store, carrying three white eggs, pausing on the outside only to look straight at Jadine and then to shoot "an arrow of saliva between her teeth down the pavement" (45–46). The woman's unsettling gesture precipitates Jadine's sudden departure for the island to escape the wealthy white man who wants her to marry him. The woman with the eggs reappears among the women of her Eloe nightmare and continues to haunt her and to threaten "the person she had worked hard to become" (262).

What Jadine has become is a woman cut off from her personal past, from her cultural heritage, and from history. Though she is emotionally vulnerable to the African woman's judgment of her choices, she refuses to act on those feelings or to take care of her aging aunt and uncle, who, along with Valerian, have paid for her education. Self-indulgent and determined to live in the present, without concern for the welfare of others, Jadine Childs is a jaded child of the post-civil-rights-movement seventies.

Intensifying the antagonisms created by race and class is a menacing threat and memory of violence. From little Pecola Breedlove in *The Bluest*

Eye to the ghost in *Beloved*, the lives of Morrison's characters are defined by acts of violence. Walking along the road after jumping ship, Son recalls the "old dread of mines" and has to remind himself that he is not in Vietnam (134). Later when he learns that a friend has died in an explosion, he cries "like an infant for all the blowings up in Asia" (225). Early in the novel, Thérèse predicts that Sydney and Ondine will kill Son: "Kill him. . . . Kill him dead" (112). Shortly after Son is found hiding in the house, Sydney, who brings him downstairs at gunpoint and patrols the halls with a gun, warns him: "If this were my house, you would have a bullet in your head" (162).

Years before he meets Jadine, Son murdered his wife and injured her teenaged lover, and when he returns with Jadine to Eloe after eight years of wandering to evade the law, he learns that his murdered wife's mother, who "slept with a shotgun every night waitin" for Son, has only recently died. Jadine and Son engage in verbal and physical fights: she threatens to kill him, and during their most serious fight, he knocks her unconscious. At the end of the novel when Son has returned to the island to find Jadine, Sydney is still waiting with a gun, swearing to put a "bullet in him for sure" (284). All his adult life Son has been stalked by those intent on killing him: in Vietnam, in his own hometown, and now on a strange island in the Caribbean.

Tar Baby is Morrison's only novel set in the post-civil-rights-movement days, approximately in the time of its publication. Jadine and Son both came of age after the Civil Rights Acts of 1964 and 1965. Son, who is now around thirty, left home at eighteen in the late sixties to join the army and fight in Vietnam. Jadine, now twenty-five, went to college sometime in the early seventies. Neither seems to appreciate the struggle that has made it possible for them to live as they do: to check into the New York Hilton, to move in and out of the white world, to contemplate—as Son does—attending college or law school, and to rent apartments in any part of New York City.

The opening events of the novel take place around 1979, though Morrison is more vague about the dating than in any of her previous novels, and readers have to add and subtract from the few given dates to come up with a time that is certainly in the late seventies. Valerian, now seventy, was thirty-nine when he met Margaret in the late forties. Their son, now thirty, the same age as Son, was born about a year after the marriage. Gideon returns from the states to his island home in 1973; some time passes, two or three years, before he begins work as the Streets' yardman, a job he keeps for three years. Son leaves his hometown in Florida in 1971 to escape the law; he has been traveling for eight years.

The year 1979 was a time when many Americans were waiting: for the energy crisis to be resolved, for the Iranian hostages to be returned, for a presidential election—or for some new violence to erupt. The overt violence of the civil rights movement and of the Vietnam War were in the past and the aftermath of Watergate had left former activists feeling vindicated, but unsure of what to do next. It was a time when those who had once focused on the atrocities of racial violence in the American South and the massacres of women and children in Southeast Asia could afford to turn their energies to the less sensational violence being done to the environment; when conservationists bought full-page ads in the *New York Times* complete with pictures depicting the murder of baby seals; and when the absence of war and civil unrest made possible the retreat into a kind of personal isolationism. It was a time of lulls, of in-betweens. It was also a time when young people could live with little awareness of the outside world in a way that was not possible in 1919 (*Sula*), 1941 (*The Bluest Eye*), or 1963 (*Song of Solomon*).

Brooding over the lives of the characters in this novel is a sense of unreality, of living in suspended time, as each is waiting for something to happen. Margaret, and with her the whole household, is waiting for the arrival of Michael, who like Godot never comes; Valerian waits for the mail; Son waits for Jadine to join him in New York; Jadine and Son for some sign about what they should do with their lives; Sydney and Ondine for the ax of unemployment to fall; and Alma Esté for Son to bring her the wig she longs for. The characters are held in limbo by their indifference to the outside world and their preoccupation with the unfinished business of the past: Margaret is tied to her memories of herself as an abusive parent, while insisting that she is "not one of those women in the *National Inquirer*" (209); Valerian is haunted by images of his father's death, his tortured two-year-old son, and his first wife; Jadine is bound by her visions of chiding women condemning her choices; and Son, by memories of the Vietnam War, the wife he killed, and his romantic illusions about how life was in Eloe.

When *Tar Baby* was published in 1981, readers knew that the relatively benign period of the Carter years had given way to an administration committed to facilitating the accumulation of private wealth with little concern about the public cost. In this context, Morrison's exposure of the covert violence and exploitation that is underneath self-indulgent, privileged private lives would seem appropriate, if not prophetic. To Son, Valerian is one of the world's exploiters. When Valerian dismisses two household servants for "stealing" apples, Son muses on Valerian and the corruption of the system:

> He had been able to dismiss with a flutter of the fingers the people whose sugar and cocoa had allowed him to grow old in regal comfort . . . but he turned it into candy, the invention of which really was child's play, and sold it to other children and made a fortune in order to move near, but not in the midst of, the jungle where the sugar came from and build a palace with more of their labor and then hire them to do more of the work he was not capable of and pay them again according to some scale of value that would outrage Satan himself . . . and he probably thought he was a law-abiding man, they all did . . . especially the Americans who were the worst. (203)

Son concludes that Valerian, Margaret, the servants, and Jadine have allied themselves to "one of the killers of the world" (204).

When Jadine and Son are struggling with conflicting values, they both imagine that they are trying to rescue the other from destructive cultural elements: "He thought he was rescuing her from Valerian, meaning *them*, the aliens, the people who in a mere three hundred years had killed a world millions of years old. From Micronesia to Liverpool, from Kentucky to Dresden, they killed everything they touched including their own coastlines, their own hills and forests" (269). Jadine, on the other hand, imagines that she is protecting him from "the night women who wanted him for themselves" (269).

The struggle between Son and Jadine is emblematic of the condition of blacks and whites in the late seventies and eighties. The official barriers to an integrated society having been eliminated, the question remains as to whether a meaningful marriage of cultures can take place. Like *Song of Solomon*, *Tar Baby* ends in ambiguity: Jadine flying—first class in an airplane—off to her life as an expatriate in Paris; Son, a castaway, running "lickety-split" into the rain forest where naked, blind men race "horses like angels all over the hills" (306). Waiting for Jadine in Paris is the white man who has given her the baby seal coat; waiting for Son with a gun on the other side of the island is Sydney, who has vowed to kill him. One is heading for a self-indulgent life that depends on human and environmental exploitation, the other to the unknown darkness of the blind horsemen or perhaps to sudden death.

Neither Jadine's forsaking of responsibility for Ondine and Sydney nor Son's impulsive plunge into the jungle is a satisfactory conclusion to their struggle, and the reader is left to complete the various scenarios that their actions might create. The novel ends, then, with the tension of unfinishedness, of lives suspended in time and space, waiting for the creative imagination to spin out the future. Most readers must wish for some kind of reunion between these powerful individuals, a reunion that will metaphorically represent the longed-for peace among the warring factions of American culture, particularly the battle between the integrationists and separatists that Jadine and Son represent. But the novel, far from spelling out any resolution to the conflicts it presents, forces readers not only to spin out scenarios for the characters' personal lives but to consider what actions must be taken and changes made before any lasting peace is achieved in the public or the private arena.

In a common version of the folktale that gives the novel its title, Brer Bear uses the Tar Baby to trap Brer Rabbit, who has been stealing bottles of water from the well the other animals dug. When Rabbit encounters the Tar Baby, he gets hopelessly stuck; the other animals find him there and argue about what to do with him. In the meantime, Rabbit tricks them into throwing him into the brier patch, the one place he wants to be.

Problems arise for readers who seek a simple correlation of the novel to the tale. At first it seems easy: Son is Brer Rabbit, who comes out at night to steal water in bottles; he gets stuck in his attraction to Jadine, who is therefore Tar Baby. So Jadine is first equated with the Tar Baby and then with the Brier Patch, and the parallels to the familiar version of the fable break down. Jadine—Tar Baby—leaves Son, and he is taken to the far side of the island where he runs "lickety-split" (305) into the island's equivalent of a brier patch, no longer Jadine, but a primitive forest, where mythical naked, blind men wait for him. The details of the novel do not finally match those of the commonly known version of the folktale. Jadine's whiteness, both in skin color and affinity to the white race, are emphasized so that she scarcely resembles a *tar* baby; Son does not escape from her, but rather temporarily leaves their apartment only to return and find that she has left him. Son does not escape Jadine by using his wit, as Rabbit escapes the Tar Baby. Rather, he seems to lose her because he cannot adapt to her world or convince her to accept his.

The passages that include references to tar further contribute to the elusiveness of the title. Early in the novel, when he spies on the sleeping Jadine, Son breathes "into her the smell of tar and its shiny consistency" (120). So now it is Son who is the Tar Baby. Later, Jadine wanders into a swampy bog and gets trapped in a substance that "looks like pitch" (185), and much later when she returns to the island from New York and passes the place her legs burn "with the memory of tar" (276). So now it is she who is Brer Rabbit. The women who haunt Jadine are associated with tar. The African woman in the yellow dress has "skin like tar" (45). Thérèse observes that the legendary swamp women Jadine imagines look down at her from the trees have "a pitchlike smell," (105), and in Eloe, the night women appear to Jadine in a "pitch-dark room" (258). Later Jadine attempts to confine the women "to the brier patch" (288), forcing them into the role of Brer Rabbit.

That there are hundreds of versions of the Tar Baby tale makes it tempting to assume that if we could find the "right" one the details of the tale and of Morrison's novel would match perfectly. Many members of the audience to which she was writing are familiar with the Joel Chandler Harris version, but its details do not definitively gloss the novel. It is possible, of course, that Morrison is jiving the white reader with the false lead of the white man's version—the water bottles, etc.—but it seems more likely that this is still another example of Morrison's characteristic ambivalence, epitomized by Milkman poised on a rock, here embedded in the allusion to the folktale of the Tar Baby. By associating the Tar Baby with both Jadine and Son, the conflict between them and their values is meaningfully reinforced. The dark, more African Son, the women in the dream, the jungle-brier patch into which he disappears, and Eloe define a black African separatist position. The light-skinned Jadine, New York, and Paris suggest an American integrationist stance. That Son and Jadine are both the Tar Baby means it is uncertain which course will or should be pursued, which will fly in the 1980s and after. The two positions are as opposed and as dialectically joined as the two terms *African American.*

AFTERWORD

In 1968, Alice Walker published her first book, a slim volume of poetry entitled *Once.* The epigraph, a quotation from Albert Camus's *De l'envers et l'endroit,* might suggest a way of reading Walker's latest work, as well as those early verses: "Poverty was not a calamity for me. It was always balanced by the richness of light . . . circumstances helped me. To correct a natural indifference I was placed half-way between misery and the sun. Misery kept me from believing that all was well under the sun, and the sun taught me that history wasn't everything."[1]

Walker's *The Temple of My Familiar* (1989) strives to achieve that balance between the dark and light. A far-ranging narrative, *Temple* at first glance seems to have more in common with the novels of García Marquez than with Walker's earlier novels. Its complex narrative structure merges folk history, myth, fable, and high fantasy with a recognizable contemporary story complete with hot tubs, psychotherapists, and references to public figures. If in *The Color Purple* she moves her characters out of the darkness of personal misery into the light of personal love, in *Temple* she balances the darkness of the world's misery with the light of the sun, not by moving from one to the other but by holding her characters in a state of tension created by living consciously in historical, geographical, psychological, sexual, and spiritual domains. Her characters all grow, develop, and learn; but rather than ending in stasis and completion, or anything like "happy ever after," they are left with much work to do in the world and with, at best, tentative personal commitments. Late in the novel, when one character observes that "there are songs that people want you to sing today

. . . that are just inappropriate to the times" (383), Walker may be suggesting that *Temple*, while appropriate to the times, may not be popular with those who want happier songs and who are unprepared to confront the misery of what she has called times like these, "a lull in political protest."[2]

Approximately twice the length of her previous novels, *Temple* includes more than forty named contemporary characters in addition to those—both animal and human—who make up the world of one character's previous lives. Most of the characters are associated with two couples who live in California: Carlotta, a former college teacher and her husband, Arveyda, a rock star; and Fanny and Suwelo, once university professors and colleagues of Carlotta, who become disillusioned and eventually leave academia. Carlotta comes to California as a refugee from Latin America with her mother, Zedé, who makes elaborate costumes from tropical feathers. When we first meet Suwelo, he is traveling by train to Baltimore to the funeral of his Uncle Rafe, who has left him a house and twenty-six thousand dollars. Back in California is Fanny, the granddaughter of Celie and the daughter of Olivia of *The Color Purple.* Once Suwelo's wife, now his sometime lover and friend, Fanny is engaged in a quest for her sanity that has led her to Africa, where for the first time she meets Ola, her father—revolutionary, playwright, and minister of culture—and her half-sister, Nzingha, also a playwright. Fanny was conceived during a brief affair Olivia had with Ola shortly before she left Africa; Nzingha is the child of his marriage to an illiterate bush woman. Ola's associates include the real-life novelist Bessie Head and the presumably fictional Mary Jane Briden, a white woman—playwright, painter, and educator—whom he married so that she would not have to leave Africa when the black regime drove out the whites.

Unlike *Grange*, with its omniscient controlling narrator, *Meridian*, with its idealized heroine, or *Purple*, with its controlling vision expressed by Shug Avery and ultimately adopted by the other characters, *Temple* evolves from the interaction of many voices and withholds from readers the comfort of a single resting place—a point of view, a focus, or harmony of diverse and often contradictory voices. When the novel ends, the questing characters are in a sense just beginning to see, and the warriors for justice are preparing for the next wave of action. While *Purple*, written in the first of the Reagan years, seems to validate lives of rest and ease lived entirely in the personal sphere, *Temple*, completed in the late eighties, suggests in Ola's last words, "how endless struggle is" (347).

Sorting out the chronology of *Temple* requires effort—Walker does not build in the detailed temporal structure typical of Morrison's novels—but it is clear that the primary action takes place in the 1980s. Yuppies have moved into Rafe's neighborhood, Ola talks of Elvis Presley in the past tense, Suwelo thinks of how long it has been since he and Fanny, "Hippies at heart," married "barefoot, in the spring, underneath blossoming apple trees" (382). Arveyda, Carlotta, Suwelo, and Fanny seem to be of Walker's generation—middle-aged, looking back at their youth in the sixties. The past emerges, as characters tell each other their stories or as they read diaries, letters, and even jottings in books. While waiting to settle his uncle's estate, Suwelo spends many hours getting to know his uncle's closest friends, among them Mr. Hal and his companion, Miss Lissie, with whom Rafe has lived in a very compatible ménage à trois. Hal and Lissie sit with Suwelo and tell tales of their life together and of Lissie's many past lives; Suwelo relates to them stories of his marriage to Fanny and of his recent affair with Carlotta while Fanny was in Africa. Interspersed with these stories are letters that Fanny writes to Suwelo from her two trips to Africa; Zedé's stories of her young Indian lover, Carlotta's father; accounts of Fanny's therapy sessions; Arveyda's stories of his family; and Olivia's narrative of her youth in Africa and her new life in America.

Though there is no reference to specific events that would fix the narrative in a particular year, even the older characters are concerned with issues of the 1980s. Talking to Suwelo, Lissie warns:

> You take the way things are going in the world today. You have your poisoned rivers and your poisoned air and your children turning into critters before your eyes. You have your leaders that look like empty cartons and the politicians who look drugged. You have a world that scares everybody to death. You can't go nowhere. You can't eat anything. You can't even hardly make love. And that's just today. . . . [I'm] afraid of what living is going to look like and be like next time I come. (190–91)

Industrial pollution, the drug problem, cardboard—if not teflon—presidents, the nuclear threat, and Acquired Immunodeficiency Syndrome (AIDS) presumably unsettle the already-troubled lives of the characters, but these issues remain beneath the surface of the narrative. On April 8, 1989, speaking in Washington in support of the National March for Women's Equality and Women's Lives, just days before *The Temple of My Familiar* appeared in bookstores, Alice Walker was much more specific about "the way things are going in the world today": "Let us look around us. . . . Let us consider the depletion of the ozone; let us consider homelessness and the nuclear peril; let us consider the destruction of the rain forests—in the name of the almighty hamburger. Let us consider the poisoned apples and the poisoned water and the poisoned air and the poisoned earth."[3]

But neither in this public statement as an activist nor in the less direct context of her latest novel does Walker pose a specific political solution to the glaring social ills she addresses. While calling for the "white, male law-giver"[4] to step aside, confess his guilt in the rape of the planet and its dark-skinned people, and to desist in colluding with the forces that oppress others, Walker also suggests the necessity of a harmonious coalition to confront the white male power establishment.

For Fanny and Suwelo, at least, the civil rights movement continues to hover over the present. Suwelo even uses a phrase reminiscent of one attributed to Stokely Carmichael in 1964. Musing that "his generation of men had failed women," Suwelo recalls, "For all their activism and political development during the sixties, all their understanding of the pervasiveness of oppression, for most men, the preferred place for women had remained the home; the preferred position for women, wherever they were, supine" (28–29).[5]

Balancing the novel's ongoing concern with the oppression of women, however, is its emphasis on the growth and development of the male characters, particularly the transformation of Suwelo. As a professor of history in a California university, Suwelo has been content to teach traditional mainstream textbook history with a subtle African-American tint:

> He wanted American history, the stuff he taught, to forever be the center of everyone's attention. What a few white men wanted, thought, and did. For he liked the way he could sneak in some black men's faces later on down the line. And then trace those backward until they appeared even before Columbus. It was like a back-stitch in knitting, he imagined, the kind of history teaching that he did, knitting all the pieces, parts, and colors that had been omitted from the original design. But now to have to consider African women writers and Kalahari Bushmen! It seemed a bit much. (179)

As the recipient of stories from others, Suwelo gradually changes from a conventional professor to a willing student of quite unconventional historical texts, including Fanny's letters from Africa; Uncle Rafe's papers and books and their endless marginalia; Hal's personal history; Lissie's oral, written, and tape-recorded stories; and even the visual texts that Lissie and Hal create with their paints. Instead of the mainstream white man's history knitted together with his own carefully chosen pieces of black history, Suwelo finds himself enchanted by myth, fable, folk stories, personal history, and even the narratives incorporated in paintings.

Lissie tells Suwelo fabulous stories about her many reincarnations: how she was captured, sold into slavery, raped, and, after running away, captured again (61–71); how she lived in peace and harmony with her apelike cousins in the trees, only to have that life disrupted by violence (83–88); how she was once a white man (355–62) and at another time a lion living as the familiar of women (364–66). In life after life, Lissie recalls being sometimes the victim and sometimes the perpetrator of oppression, periodically evoking the horrors of slavery. After having been held in thrall for some months by Lissie's stories, Suwelo at last returns to California, prepared to write an oral history based on the very unconventional material that is the stuff of Hal and Lissie's lives.

More than the other characters, Fanny experiences personal distress while confronting the way the world is going and her own anger at the white people who are at fault. She recalls being traumatized by the violence during the civil rights movement: "In high school I watched the integration of the University of Georgia on television . . . the night the whole campus seemed to go up in flames, and white people raged. . . . I saw the Freedom Riders, black and white, beaten up in Mississippi. . . . I saw a lot of black people and their white allies humiliated, brutally beaten, or murdered" (298).

Fanny is terrified of the future—or the thought that there may not be a future—and like Velma in *The Salt Eaters*, she is exhausted by her life as an activist. "I've marched so much by now and been arrested so many times, I'm really quite weary" (302). Most debilitating of all is her hatred of whites and the violent impluses she feels. Freedom from the bondage of her feelings comes when, with the help of her therapist, Fanny confronts the origins of her hatred and fear of whites. Her now middle-aged and only white childhood friend, Tanya, whose story might have been included in *Meridian*, explains that she, too, was changed by the movement, in that she married a black man, a "political shortcut" that allowed her to feel that she was doing something about racism without changing society. When Fanny asks how she changed from the way she was raised, Tanya explains: "The Civil Rights Movement happened. Selma happened. The University of Georgia happened. Dr. King happened. It just hit me one night, watching television coverage of one of the Civil Rights marches, that the order of the world as I'd always known it, and imagined it would be forevermore, was *wrong*" (326).

Through Tanya's stories, Fanny reclaims long-repressed memories of growing up black in a racist world. Traveling to Africa and discovering her father, Ola, the most political of the characters—contributes to her recovery. In Ola's country, as in the post-civil-rights-movement United States, oppression is widespread. More than once Ola expresses his conviction that "the present you are constructing . . . should be the future you want" (236), and he does not hesitate to give his daughter specific advice: "To the extent that it is possible . . . you must live in the world today as you wish everyone to live in the world to come. That can be your contribution" (336). Explaining to her how Mary Jane Briden and her staff created a school that made a difference for the children and their country's future, Ola observes that they had a vision of the future that "looked an awful lot like what they already had together every day" (345). In times like these, lulls perhaps, Ola's observations suggest that the very least those committed to social change can do is to design their personal lives as models for the world they hope to create for others. The model life, however, is not necessarily an idyllic one, and it may include violence.

In her fiction Walker repeatedly raises the question of violence and its place in personal and social transformation. The apparently necessary murder of Brownfield at the end of *The Third Life of Grange Copeland* suggests that in some cases violence against oppressive forces may be the only way to free the oppressed. In *Meridian*, however, the role of violence remains problematic, as Meridian herself raises but never definitively answers questions about whether she could "kill for the Revolution." In *The Color Purple* the chain of personal violence is broken only when Shug convinces Celie not to kill Albert. But *The Temple of My Familiar* once again raises the issue of violence as Ola and Olivia debate with their daughter Fanny about whether violence has a place in the struggle for justice. Ola argues that the fight to overthrow the white regime in his country required ruthlessness and violence and that for him there "seemed no other way" (305). Fanny's mother, on the other hand, counsels against violence and advocates forgiveness: "Forgiveness is the true foundation of health and happiness, just as it is for any lasting progress. Without forgiveness there is no forgetfulness of evil; without forgetfulness there still remains the threat of violence. And violence does not solve anything; it only prolongs itself" (308).

Fanny, however, does not entirely support either position. She wonders if "perhaps" her father might be right in choosing a violent fight for freedom in Africa, but she cannot see how his position could be applied to

1980s America: "In the United States there is the maddening illusion of freedom without the substance. It's never solid, unequivocal, irrevocable. So much depends on the horrid politicians the white majority elects. Black people have the oddest feeling, I think, of forever running in place" (305–06). Equally unrealistic to Fanny is her mother's position of nonviolence and forgiveness: "The way things are going in the United States . . . there will soon be more black men in prison than on the streets. In South Africa the entire black population is incarcerated in ghettos and 'homelands' they despise. . . . Forgiveness isn't large enough to cover the crime" (308).

Further complicating the characters' search for appropriate action in a world dominated by the greedy and self-seeking is the possibility that all could be lost in a nuclear holocaust. Demonstrating considerable knowledge of nuclear winter and its accompanying horrors, Mr. Hal observes that "if they're going to blow us up, or make us freeze to death and starve in the dark, we might as well be enjoying ourselves by having a good story" (274). But Fanny cannot help but be afraid. Responding to her fear of an accidental holocaust, Suwelo insists that the whole world is being affected by the nuclear threat: "Prior to this time in history, at least we thought we'd have a future, that our children would see freedom, even if we never did. Now they've made sure that none of our children will ever live the free and healthy lives so many generations of oppressed people have dreamed of for them. And fought so hard for. I very often think of violence, but any violence I could do at this point would seem, and be, so small" (302).

The debate between those who advocate violent revolution and those who adhere to nonviolence that divided movement activists in the sixties is still alive in Walker's fiction, and though Fanny does not resolve this or what Ola calls the other "eternal questions" the novel raises, she does confront her own racism, violent impulses, and fear of whites; she progresses a little further down the road toward the inner harmony both her father and mother insist is necessary for progress. Olivia argues that neither revolution nor personal transformation is fostered by those who are fighting within themselves, but rather by those who, "unbroken, uncorrupted," die "with the same passion with which they'd lived," and who in the end appear "to see . . . the beloved community of souls with whom they'd kept the faith" (310). Similarly, Ola urges their daughter to "harmonize" her heart: "Only you will know how you can do that; for each of us it is different. Then harmonize, as much as this is ever possible, your surroundings" (316).

While the novel does not provide any prescription for achieving personal growth and inner harmony, it does, through the stories of individual characters, affirm the possibility of achieving balance, sanity, and health, and it suggests that social progress is best achieved by those engaged in personal growth. The "how to" of personal and public change remains individual and problematic. And indeed, Fanny, Suwelo, Carlotta, and Arveyda seem to travel on independent paths.

Most of the characters are or become artists in the course of the novel. Ola, Mary Jane, Nzingha, and Fanny are playwrights; Carlotta and Arveyda compose music; Zedé and her mother are feather artists; and Hal, Lissie, Mary Jane, and even Ola's illiterate wife are painters. There are also a furniture maker, a photographer, and two novelists. Woven into the narrative are suggestions that in times like these, after the revolution and before it is clear what the next social movement might be, artists and writers can be the most effective agents for social change. Ola's African daughter, who argues that writers "cause as much trouble as anyone," explains her father's views: "Writers don't cause trouble so much as they describe it. Once it is described, trouble takes on a life visible to all, whereas until it is described, and made visible, only a few are able to see it" (259).

At the end of the novel, after having regained a certain personal harmony, Fanny has found a way to combine social activism with creative work. She is preparing to go back to Africa to produce a play she is writing with her sister, and though she believes there is a good chance she will be arrested, she is prepared to do her part to insure that Africa will "belong to all its people, the women as well as the men" (389). Fanny's play will incorporate scenes from three of her father's plays, and will presumably be a continuation of his work.

On his last visit to his friends, Suwelo falls asleep on the sunny porch of their house. When he wakes he realizes Lissie and Hal have painted him surrounded by flowers and fruit trees, "all the things they loved" (193). This painting of Suwelo, a sun-drenched counter to the dark hold of the slave ship that Lissie has also evoked, provides the balance on which Walker has always insisted in her poetry and fiction.

The content of the rich body of novels written by African-American women in the wake of the civil rights movement includes that dark hold and the sun-drenched scene and almost everything in between. In slightly less than a quarter of a century black women writers have produced a substantial and significant body of novels, some among the most compelling of our times. By confronting in a fictional context the question—Where do we go from here?—that Martin Luther King, Jr., was publicly asking at the time of his death and that his followers, as well as his opponents, have continued to ask, black women novelists may have answered the question in the process of asking it. Where they are likely to go is into increasingly challenging imaginative worlds where the thorniest issues of race, class, and gender can be wrestled with and brought forth into the public discourse. Their sensitivity to what others can hear and a determination to speak from and into the times has resulted in an amazing variety of voices making trouble "visible," asking the most troubling questions—and being heard.

NOTES

Introduction

1 For a succinct discussion of major positions within the discourse of racial justice see the introductions to the chapters in August Meier, Elliott Rudwick, and Francis L. Broderick, *Black Protest Thought in the Twentieth Century*, 2d ed. (Indianapolis: Bobbs-Merrill, 1971). Discussions of events preceding the movement include C. Vann Woodward, *The Strange Career of Jim Crow*, 3d rev. ed. (New York: Oxford University Press, 1974), and Mary Frances Berry and John W. Blassingame, *Long Memory: The Black Experience in America* (New York: Oxford University Press, 1982). For overviews of the movement see Harvard Sitkoff, *The Struggle for Black Equality, 1954–1980* (New York: Hill and Wang, 1981), Jack M. Bloom, *Class, Race, and the Civil Rights Movement* (Bloomington: Indiana University Press, 1987), and Robert Weisbrot, *Freedom Bound: A History of America's Civil Rights Movement* (New York: Norton, 1990). For chronological accounts of movement highlights see Juan Williams, *Eyes on the Prize: America's Civil Rights Years, 1954–1965* (New York: Viking, 1987), the accompanying volume for the Public Broadcasting Service series of the same title. See also the companion volume to the second part of that series: Henry Hampton and Steve Fayer, *Voices of Freedom: An Oral History of the Civil Rights Movement from the 1950s through the 1980s* (New York: Bantam, 1990). See also Clayborne Carson, *In Struggle: SNCC and the Black Awakening of the 1960s* (Cambridge: Harvard University Press, 1981).

2 Christian, "But What Do We Think We're Doing Anyway," in *Changing Our Own Words: Essays on Criticism, Theory, and Writing by Black Women*, ed. Cheryl A. Wall (New Brunswick: Rutgers University Press, 1989), 67.

3 Hazel Carby, *Reconstructing Womanhood: The Emergence of the Afro-American Woman Novelist* (New York: Oxford, 1987); Barbara Smith, "Toward a Black Feminist Criticism," *Conditions: Two* 1 (October 1977); Deborah McDowell, "New Directions for Black Feminist Criticism," *Black American Literature Forum* 14 (October 1980); Barbara Christian, *Black Feminist Criticism: Perspectives on Black Women Writers* (New York: Pergamon, 1985).

4 Carby, *Reconstructing Womanhood*, 7.

5 Christian, "But What Do We Think," 72.

6 Henry Louis Gates, Jr., *The Signifying Monkey: A Theory of Afro-American Criticism* (New York: Oxford University Press, 1988).

7 Marjorie Pryse and Hortense J. Spillers, *Conjuring: Black Women, Fiction, and the Literary Tradition* (Bloomington: Indiana University Press, 1985), 21, 258.

8 Toni Cade, ed., *The Black Woman* (New York: New American Library, 1970), 11.

9 See Mary Helen Washington, ed., *Black-eyed Susans: Classic Stories by and about Black Women* (New York: Anchor, 1975), xi. Also relevant to this study are Washington, ed., *Midnight Birds: Stories of Contemporary Black Women Writers* (New York: Anchor, 1980), and Washington, ed., *Invented Lives: Narratives of Black Women, 1860–1960* (New York: Anchor, 1987).

10 Other relevant works include Barbara Christian, *Black Women Novelists: The Development of a Tradition* (Westport: Greenwood Press, 1980); Claudia Tate, ed., *Black Women Writers at Work* (New York: Continuum, 1983); Gloria Wade-Gayles's *No Crystal Stair: Visions of Race and Sex in Black Women's Fiction* (1984), which addresses issues of race and gender in some twelve novels published between 1946 and 1976; Mari Evans's, ed., *Black Women Writers, 1950–1980: A Critical Evaluation* (Garden City, New York: Anchor, 1984). Houston A. Baker, Jr., and Joe Weixlmann, eds., *Black Feminist Criticism and Critical Theory* (1988); and some essays included in Joanne M. Braxton and Andrée Nicola McLaughlin, eds., *Wild Women in the Whirlwind: Afra-American Culture and the Contemporary Literary Renaissance* (1990). Other works that are at least in part concerned with contemporary black women writers include Melvin Dixon, *Ride Out the Wilderness* (Urbana: University of Illinois Press, 1987), Calvin Hernton, *The Sexual Mountain and Black Women Writers* (New York: Anchor, 1987), and Charles Johnson, *Being and Race: Black Writing since 1970* (Bloomington: Indiana University Press, 1988), 94–118. Journal issues devoted to black women writers include *Black Scholar* 17 (March–April 1986). For a collection of reviews, interviews, and essays concerning Toni Morrison, see Nellie McKay, *Critical Essays on Toni Morrison* (Boston: G. K. Hall, 1988). Clyde Taylor's "Black Writing as Immanent Humanism," in *Afro-American Writing Today: An Anniversary Issue of the Southern Review* [1985], ed. James Olney (Baton Rouge: Louisiana State University Press, 1989), 203–13.

11 Norman Harris's *Connecting Times: The Sixties in Afro-American Fiction* (Jackson: University Press of Mississippi, 1988) addresses the relation between events

in the sixties and early seventies and selected texts—all published between 1971 and 1978. Though he includes chapters that deal with the civil rights and black power movements, Harris treats only one novel by a black woman, Alice Walker's *Meridian.* In *Specifying: Black Women Writing the American Experience* (Madison: University of Wisconsin Press, 1987), Susan Willis considers some historical and cultural implications of texts by Zora Neale Hurston, Paule Marshall, Toni Morrison, Alice Walker, and Toni Cade Bambara.

Chapter 1: Slavery and Reconstruction

1 Margaret Walker, *How I Wrote Jubilee and Other Essays on Life and Literature,* ed. Maryemma Graham (New York: Feminist Press, 1990), 53–56.
2 Interview, National Public Radio, October 1, 1987. For further elaboration of *Beloved* as a memorial, see Robert Richardson's interview with Morrison in *The World,* 3 (January–February 1989), 4.
3 *How I Wrote Jubilee,* 57.
4 Juan Williams, *Eyes on the Prize,* 283–84.
5 *How I Wrote Jubilee,* 61.
6 See Hazel Carby, "Ideologies of Black Folk: The Historical Novel of Slavery," in *Slavery and the Literary Imagination: Selected Papers of the English Institute, 1987* (Baltimore: John Hopkins University Press, 1989), 129, 136. Concerned with how historical fiction is "generated from particular cultural conditions," and recognizing that *Jubilee* engages "directly the concerns of the civil rights movement," Carby concludes that Walker's novel in 1966 offered "a severely limited historical, psychological, and aesthetic vision of the possibilities of a free black community." *Jubilee* is briefly mentioned and compared with *Gone with the Wind* in Roger Whitlow, *Black American Literature* (Chicago: Nelson-Hall, 1973; reprint, Totowa, N.J.: Littlefield, Adams, 1974), 138–39. For a different approach that combines a consideration of *Jubilee* with Hurston's *Their Eyes Were Watching God* and Morrison's *Sula* see Hortense J. Spillers, "A Hateful Passion, A Lost Love," *Feminist Studies* 9 (Summer 1983), 293–323. Spillers places *Jubilee*'s historical dimension in a large context dictated by "Divine Will," 299.
7 For a discussion of the mulatto figure as "a narrative device of mediation," see Carby, *Reconstructing Womanhood,* 88–91.
8 Minrose Gwin in "Jubilee: The Black Woman's Celebration of Human Community," *Conjuring,* assumes that the novel, through Vyry's voice, proposes "black humanism as an answer to America's racial conflicts" (148). In an interview published in 1979, Margaret Walker explains that many readers misunderstand *Jubilee,* that she did not mean for Vyry's position to be the voice of the novel, rather that she "tried to show several points of view in *Jubilee*" (*Callaloo* 2 [May 1979], 291).

9 See Trudier Harris, "Black Writers in a Changed Landscape, since 1950," in Louis D. Rubin, Jr., et al., *The History of Southern Literature* (Louisiana State University Press, 1985), 566–77. Harris concludes that *Jubilee* was "outdated from the date of its publication" and that it has failed to inspire younger writers who objected to its loyalty to the heritage of bondage "that blacks were still struggling to overcome in the 1960s." Deborah McDowell, on the other hand, considers that *Jubilee* may have been the catalyst for numerous novels about slavery published in the last two decades, including Ernest Gaines's *The Autobiography of Miss Jane Pittman* (1971), Ishmael Reed's *Flight to Canada* (1976), Barbara Chase-Riboud's *Sally Hemmings* (1979), Octavia Butler's *Kindred* (1979), Charles Johnson's *The Oxherding Tale* (1982), as well as Sherley Anne Williams's *Dessa Rose* and Toni Morrison's *Beloved.* See Deborah McDowell, "Negotiating Between Tenses: Witnessing Slavery after Freedom—*Dessa Rose*," in Deborah E. McDowell and Arnold Rampersand, eds., *Slavery and the Literary Imagination: Selected Papers of the English Institute* (Baltimore: Johns Hopkins University Press, 1989), 125–63. Chase-Riboud's latest novel, *Echo of Lions* (1989), is also about a slave revolt.

10 See Washington, *Midnight Birds*, 248.

11 For the story of the pregnant rebel slave woman who was permitted to remain in jail until after she gave birth, see Herbert Aptheker, *American Negro Slave Revolts* (New York: Columbia University Press, 1943; reprint, New York: International Publishers, 1963), 287. Reference to the white woman "living in a very retired situation" who aided runaways is found on page 289.

12 See Vincent Harding, *There Is a River: The Black Struggle for Freedom in America* (New York: Harcourt Brace Jovanovich, 1981).

13 The story of Margaret Garner, the woman on whom Sethe's character is based, is told in Herbert Aptheker's "The Negro Woman," *Masses and Mainstream* 11 (February 1948), 11, and in Angela Davis's *Women, Race and Class* (New York: Random House, 1981), 21. A facsimile of a newspaper account of a visit to Garner while she was in jail relating the details of the murder as Morrison presents them is included in Middleton Harris, et al., *The Black Book* (New York: Random House, 1974), 10. Morrison edited both Davis and Harris.

14 See Rom. 9:25, Hos. 1 and 2.

15 Elizabeth Kastor, "Toni Morrison's 'Beloved' Country," *Washington Post*, October 5, 1987, B12.

Chapter 2: From the Great War to World War II

1 Bernard Bell, *The Afro-American Novel and Its Tradition* (Amherst: University of Massachusetts Press, 1987), 263.

2 For a discussion of rape in these and other African-American novels, see Missy Dehn Kubitschek, "Subjugated Knowledge: Toward a Femininist Exploration

of Rape in Afro-American Fiction," in Baker and Weixlmann, *Black Feminist Criticism*, 43–56.

3 Toni Morrison, in Charles Ruas, ed., *Conversations with American Writers* (New York: Alfred A. Knopf, 1985), 220. In "Unspeakable Things Unspoken: The Afro-American Presence in American Literature," *Michigan Quarterly Review* 28 (Winter 1989), 20–24, Toni Morrison reveals that *The Bluest Eye* was composed from 1965–69 and *Sula* begun in 1969. Referring to those years as a time of "great social upheaval in the life of black people," she counsels readers to consider her novels in the context of that "period of extraordinary political activity."

4 For different perspectives on *The Bluest Eye*, see Michael Awkward, "Roadblocks and Relatives: Critical Revision in Toni Morrison's *The Bluest Eye*," Caroline Denard, "The Convergence of Feminism and Ethnicity in the Fiction of Toni Morrison," and Trudier Harris, "Reconnecting Fragments: Afro-American Folk Tradition in *The Bluest Eye*," in *Critical Essays on Toni Morrison*, ed. Nellie McKay (Boston: G. K. Hall, 1988); Barbara Christian, *Black Women Novelists*, 138–53; and Phyllis R. Klotman, "Dick and Jane and the Shirley Temple Sensibility in *The Bluest Eye*," *Black American Literature Forum* 13 (Winter 1979), 123.

5 The significance of the doll test and its implications to educational policies continues to concern researchers. For a more recent study, see M. B. Spencer, "Cultural Cognition and Social Cognition of Black Children's Personal-Social Development," in G. K. Brokins and W. R. Allen, eds., *Beginnings: The Social and Affective Development of Black Children* (Hillsdale, N.J.: Earlbaum, 1985). Caroline Denard discusses Morrison's concern with the damage done by the widespread acceptance of the white aesthetic, in "The Convergence of Feminism and Ethnicity in the Fiction of Toni Morrison," in McKay, *Critical Essays*, 172.

6 Juan Williams, *Eyes on the Prize*, 23.

7 Morrison discusses the significance of the fall of 1941 to the opening scene of *The Bluest Eye* in "Unspeakable Things Unspoken: The Afro-American Presence in American Literature," *Michigan Quarterly Review* 28 (Winter 1989), 21.

8 Gloria Naylor and Toni Morrison, "A Conversation," *The Southern Review* 21 (Summer 1985), 577.

9 For Walker's version of how this novel came to be written, see "Writing *The Color Purple*" in her collection of essays *In Search of Our Mother's Gardens* (New York: Harcourt Brace Jovanovich, 1983), 355–60. Brief discussions of *The Color Purple* are included in Marjorie Pryse, "Zora Neale Hurston, Alice Walker, and the 'Ancient Power' of Black Women," in Pryse and Spillers, *Conjuring*.

10 For a discussion of the merging of voices in *The Color Purple*, see Henry Louis Gates, Jr., "Color Me Zora: Alice Walker's (Re)Writing of the Speakerly Text," in *The Signifying Monkey*, 239–58.

11 Woodward, *Strange Career of Jim Crow*, 96.

12 For the history of alternative independence day celebrations in the black community, see Berry and Blassingame, *Long Memory*, 55–56.
13 Alice Walker, *Living By the Word* (New York: Harcourt Brace Jovanovich, 1988), 92.

Chapter 3: Harbingers of Change: Harlem

1 For an overview of Harlem in the 1920s see Nathan Irvin Huggins, *Harlem Renaissance* (New York: Oxford University Press, 1971). See also Jervis Anderson, *This Was Harlem: A Cultural Portrait, 1900–1950* (New York: Farrar, Straus, Giroux, 1983.)
2 Anderson, *Harlem*, 280.
3 Huggins, *Harlem Renaissance*, 3.
4 Jervis Anderson, *A. Philip Randolph: A Biographical Portrait* (New York: Harcourt Brace Jovanovich, 1973; reprint, Berkeley: University of California Press, 1986), 5.
5 Anderson, *Randolph*, 325.
6 Anderson, *Harlem*, 61.
7 Rosa Guy, *Children of Longing* (New York: Holt, Rinehart, 1970), xiii. A collage of interviews with and essays by black youth, stitched together with Guy's commentary, this unusual volume grew from a trip she took south following the assassination of Martin Luther King, Jr., to discover the effects of the movement and its accompanying violence on the lives of young African Americans.
8 Rosa Guy, *New York Times Book Review*, May 31, 1987, 36.
9 Other Harlem novels that focus on the lottery are Julian Mayfield's *The Hit* (1957) and *The Long Night* (1958).
10 For a discussion of the role of Powell's Abyssinian Baptist Church in the political life of Harlem, see Anderson, *Harlem*, 23. It became a powerful political force in the 1930s, when Adam Clayton Powell, Sr., stepped back to let his son take the spotlight.
11 For a discussion of the two aspects of Francie's personality see Nellie McKay, "Afterword" to *Daddy Was a Number Runner* (New York: Feminist Press, 1986), 210.
12 See Rita B. Dandrige, "From Economic Insecurity to Disintegration: A Study of Character in Louise Meriwether's *Daddy Was a Number Runner*," *Negro American Literature Forum* 9 (Fall 1975), 82–85.
13 Public address, Emory University, Atlanta, Ga., April 21, 1986.
14 *A Short Walk* uses more historical material than may be apparent to readers unfamiliar with the Harlem Renaissance. For example, there was an actress named Cora Green who played a part in Harlem during its heyday; much of the Garvey material is based on documented historical events.
15 Although African Americans had made significant contributions to the military

since the American Revolution, it was not until 1948 that an American president saw fit to eliminate discrimination and inequities for black men and women in the armed services. Truman's seemingly modest and long-overdue act, however, was so radical in the minds of some that, along with eliminating other racist practices, it cost him the support of four southern states in the 1948 election. Amazingly, both congressmen and top military officers who implemented the order chose to keep the changes from the press to avoid an adverse public reaction so that, according to C. Vann Woodward, many Americans were not aware of the impact of Truman's order until 1953 (*Strange Career of Jim Crow*, 137).

16 Perhaps in part because many of her books have been classified as appropriate for young adults, Rosa Guy has not yet received the critical attention she deserves. For biographical information and a discussion of her other work, see Jerrie Norris, *Presenting Rosa Guy* (Boston: Twain, 1988).

Chapter 4: Private Lives before the Movement

1 Walker dedicates this novel to her mother, "who made a way out of no way." This phrase, from the lyrics of the traditional gospel song "Jesus will make a way out of no way," is also found in Toni Morrison's *Beloved* (95).
2 Morrison, "Unspeakable Things," 23–27.
3 Morrison, "Rootedness: The Ancestor as Foundation," in Evans, *Black Women Writers*, 339.
4 Morrison, "Rootedness," 344.

Chapter 5: From Desegregation to Voting Rights

1 Toni Morrison has said that Milkman's hometown is Flint, not Detroit as some critics have assumed (personal communication, 1984). When Milkman hitches a ride and the driver asks him if he knows Flint, Milkman says he does.
2 Odessa Farrell, Educational Director of the NAACP in St. Louis, 1989, reports that in 1959 integration was proceeding according to a busing plan set in motion in September 1954. There was no major upheaval in the city at that time. In the 1980s, however, school integration was once again an issue in St. Louis, and in 1983, a "Settlement Agreement" was handed down by the courts mandating that black students transfer to the mostly white county schools (personal communication, 1989).
3 Georg Lukacs, *The Historical Novel* (Boston: Beacon, 1963), 81.
4 For a discussion of Morrison's treatment of characters crippled by their isolation from community, see Valerie Smith, "The Quest for and Discovery of Identity in Toni Morrison's *Song of Solomon*," in *Afro-American Writing Today*, ed. James Olney, 134–45.

5 Morrison, in Evans, *Black Women Writers*, 344–45.

6 There is a discrepancy of Morrison's dating of this episode since she notes that Milkman is twenty-two, which he would have been in 1953. Emmett Till was actually killed in 1955.

7 See Dan T. Carter, *Scottsboro: A Tragedy of the American South* (Baton Rouge: Louisiana State University Press, 1969), xi.

8 Alex Haley, *The Autobiography of Malcolm X* (New York: Grove Press, 1964, 65; reprint, New York: Ballantine, 1973), 194.

9 Woodward, *Strange Career of Jim Crow*, 114–15.

10 Arthur Waskow I, *From Race Riot to Sit-in* (Garden City, N.Y.: Doubleday, 1966).

11 For a brief overview of African Americans during and following World War I, see "Military Service and the Paradox of Loyalty," in Berry and Blassingame, *Long Memory*, 314–20, and "The African American in World War I," in Herbert Aptheker, *Afro-American History: The Modern Era* (Secaucus, N.J.: Citadel Press, 1971), 159–72.

12 Paula Giddings, *When and Where I Enter* (New York: Morrow, 1984), 222.

13 See Trudier Harris, "The Barbershop in Black Literature," *BALF*, 13 (Fall 1979), 112–18.

14 Evans, *Black Women Writers*, 341.

15 In black folklore, a guitar is, according to Arna Bontemps, an instrument of the devil. See the introduction, *The Book of Negro Folklore*, ed. Langston Hughes and Arna Bontemps (New York: Dodd Mead, 1958; reprint New York: Dodd, Mead, 1983), xiii. For a discussion of Morrison's use of names in *Song of Solomon*, see Joyce Ann Joyce, "Structural and Thematic Unity in Toni Morrison's *Song of Solomon*," *CEA Critic* 49 (Winter 1987–Summer 1987), 185–97.

16 W. E. B. Du Bois, *The Souls of Black Folk* (Chicago: A. C. McClurg, 1903; reprint, New York: Fawcett, 1961), 17. Du Bois notes the black's experience of ever feeling "his twoness,—an American, a Negro; two souls, two thoughts, two unreconciled strivings; two warring ideals in one dark body, whose dogged strength alone keeps it from being torn asunder."

17 Tate, *Black Women Writers at Work*, 84.

18 Ibid., 83.

Chapter 6: In the Wake of the Movement

1 For Walker's own assessments of the movement and the role she played as a civil rights activist, see the essays in part 2 of *In Search of Our Mothers' Gardens*.

2 See Naylor and Morrison, "Conversation," 573.

3 Tate, *Black Women Writers at Work*, 176.

4 For further discussion of Walker's use of the image of quilt making, see Barbara Christian, "Alice Walker: The Black Woman Artist as Wayward," in Evans,

Black Women Writers, 457–77; Houston A. Baker, Jr., and Charlotte Pierce-Baker, "Patches: Quilts and Community," in *Afro-American Writing Today*, ed. James Olney. Barbara Christian periodically returns to the quilt metaphor in "Novels for Everyday Use," in *Black Women Novelists*, 180–238. For Walker's discussion of her own work as a quilter, see "Writing the Color Purple," in *In Search of Our Mothers' Gardens*, reprinted in Evans, *Black Women Writers*, 453–56. For another angle, see Elsa Barkley Brown, "African-American Women's Quilting: A Framework for Conceptualizing and Teaching African-American Women's History," *Signs* 14 (Summer 1989), 921–29.

5 Among the white civil rights activists who paid a personal price for their involvement in the movement was Joni Rabinowitz, a student from Antioch College (Howard Zinn, *SNCC: The New Abolitionists* [Boston: Beacon, 1965], 212).

6 *Meridian* is also dedicated to Staughton Lynd, a white historian and director of the freedom school program, and to a "Mary*am*" L.

7 Tate, *Black Women Writers*, 185.

8 Alice Walker, *In Search of Our Mother's Gardens*, 28.

9 Gloria T. Hull provides helpful guidance for reading *The Salt Eaters* in "'What It Is I Think She's Doing Anyhow': A Reading of Toni Cade Bambara's *The Salt Eaters*," in Pryse and Spillers, *Conjuring*, 216–32; see also Eleanor W. Traylor, "Music as Theme: The Jazz Mode in the Words of Toni Cade Bambara," in Evans, *Black Women Writers*, 58–70.

10 See Toni Cade Bambara, "What It Is I Think I'm Doing Anyhow," in *The Writer and Her Work*, ed. Janet Sternburg (New York: W. W. Norton, 1980), 157.

11 For Bambara's explanation of the working titles of this novel and her view of the relationship of her work to public events including the movement, see Bambara, "What It Is I Think," 156, 165, 166.

12 See Bambara, "What It Is," 164.

13 Bambara credits the "Neo-Black Arts Movement" with achieving breakthroughs in the possibilities for the novel ("What It Is," 167).

14 Kermode, *Sense of an Ending* (New York: Oxford University Press, 1967), 129.

15 For this version of the tar baby tale, see Hughes and Bontemps, *The Book of Negro Foklore*, 1–2. In Thomas LeClair, "A Conversation with Toni Morrison: 'The Language Must Not Sweat,'" *New Republic*, March 21, 1981, 27, Morrison explains that by merging the Brer Rabbit tale with the African myth of the tar lady, she was "dusting off the myth to see what it might conceal." For different interpretations of the tale of the tar baby, see Dorothy H. Lee, "The Quest for Self: Triumphs and Failure in the Works of Toni Morrison," in Evans, *Black Women Writers*, 355–56 and Angelita Reyes, "Ancient Properties in the New World: The Paradox of the 'Other' in Toni Morrison's *Tar Baby*," *The Black Scholar* 17 (March–April 1986), 19–25. Also see Trudier Harris, *Exorcising Blackness* (Bloomington: Indiana University Press, 1984), 149–62.

Afterword

1 Alice Walker, *Once: Poems* (New York: Harcourt Brace Jovanovich, 1968). Like Camus, Walker grew up in a deprived and by some standards an impoverished environment. They both had fathers who labored on the margins of a farm community; their mothers worked as domestic servants. But social and economic marginality was only one part of their common experience. Both grew up in a sunny climate—Camus in Algeria and Walker in Georgia—and found in the natural world a counter to the darkness of poverty and oppression. Like Camus, Walker combines a concern for social change with a commitment to nurturing beauty and joy in life, but for her, the range of social ills has expanded far beyond what she might have dreamed when she first encountered the work of Albert Camus, probably in the early sixties when his works were passed around and carefully read by movement activists.

2 Walker, *In Search of Our Mothers' Gardens*, 182.

3 Alice Walker, "What Can the White Man Say to the Black Woman," *The Nation*, May 22, 1989, 692.

4 Walker, "What Can the White Man Say," 692.

5 See Mary King, *Freedom Song: A Personal Story of the 1960s Civil Rights Movement* (New York: William Morrow, 1987), 452. King recalls that Carmichael's remark that "the position of women in SNCC is prone" was made in jest and received in good humor by the women present.

BIBLIOGRAPHY OF PRIMARY WORKS

The following eighteen novels are cited in the text by page numbers that refer to the editions listed here.

Bambara, Toni Cade. *The Salt Eaters.* New York: Random House, 1980. Page numbers are the same for the Vintage edition.

Childress, Alice. *A Short Walk.* New York: Avon, 1979.

Guy, Rosa. *A Measure of Time.* New York: Bantam, 1983.

Hunter, Kristin. *The Lakestown Rebellion.* New York: Charles Scribner's, 1978.

Meriwether, Louise. *Daddy Was a Number Runner.* New York: Jove Books, 1970.

Morrison, Toni. *Beloved.* New York: Knopf, 1987.

———. *The Bluest Eye.* New York: Washington Square Press, 1970.

———. *Song of Solomon.* New York: Knopf, 1977.

———. *Sula.* New York: New American Library, 1973.

———. *Tar Baby.* New York: New American Library, 1981.

Shange, Ntozake. *Betsey Brown.* New York, St. Martin's, 1985.

———. *Sassafrass, Cypress, and Indigo.* New York: St. Martin's, 1982.

Walker, Alice. *The Color Purple.* New York: Harcourt Brace Jovanovich, 1982.

———. *Meridian.* New York: Washington Square Press, 1976.

———. *The Temple of My Familiar.* New York: Harcourt Brace Jovanovich, 1989.

———. *The Third Life of Grange Copeland.* New York: Harcourt Brace Jovanovich (Harvest), 1970.

Walker, Margaret. *Jubilee.* New York: Bantam, 1966.

Williams, Sherley Anne. *Dessa Rose.* New York: William Morrow, 1986.

INDEX